Lime and Lime Mortars

A D Cowper

First published in 1927 for
the Building Research Station
by HM Stationery Office, London

Routledge
Taylor & Francis Group
LONDON AND NEW YORK

BRE

First published in 1927 for the Building Research Station
by HM Stationery Office, London

Published by Donhead Publishing Ltd 1998

Published 2015 by Routledge
2 Park Square, Milton Park, Abingdon, Oxon, 0X14 4RN
711 Third Avenue, New York, NY 10017, USA

Routledge is an imprint of the Taylor & Francis Group, an informa business

ISBN 978-1-873394-29-8 (pbk)

A CIP catalogue for this book is available from the British Library

FOREWORD TO THE 1998 EDITION

In the early years of BRE, or the BRS as it was then, a series of Special Reports and Technical Papers were produced from 1921 onwards. The Special Reports were generally review and compilations of previous research by BRS and others, whilst the Technical Papers were BRS research papers. The list of topics covered would not be out of place in a current list of research projects, and included heat transmission through walls, fire resistant construction and the use of cob in walls, as well as cement and concrete. Probably the best known of the Special Reports is No. 18 *"The Weathering of Natural Building Stones"*, by R. J. Schaffer (1932) and known almost universally as 'Schaffer'. This volume was reprinted in 1948 and a facsimile produced by HMSO in 1985 which is still quoted by many authorities as a basic textbook for the subject.

The list of authors for the other Reports and Papers includes many of the great names of building materials research including Lea, Brady, Davy, Glanville and Cowper. The Special Report by Cowper on *"Lime and Lime Mortars"* (No. 9) was originally published in 1927 and provides a comprehensive review of lime burning, the classification of limes, the use of limes, limewashes, lime plaster, pozzolanas, mortars and concretes. When first published it represented a 'state-of-the-art' review and it remained a valuable source of information and guidance for many years.

The revival in interest in the use of traditional lime mortars, plasters and washes has led to a revival in the fortunes of Cowper's Report. Its reproduction in the facsimile edition will allow the valuable information it contains to be made available to both those who have known of the report for many years and to those who are new to the study of lime mortars, plasters and washes.

Tim Yates October 1997

PREFATORY NOTE

THE present Report is not a discussion of research work carried out at the Building Research Station; it is merely a compilation of information preliminary to such work. The various tests mentioned have, however, been tried at this Station.

A summary of existing information was rendered necessary not only by the needs of the Station, but also by the number of inquiries received. It will be remembered further that a committee of the British Engineering Standards Association has recently been set up to consider the drawing up of a standard specification for lime. The suggested specifications appended to this report are only tentative, and may require modification as the result of the work of the British Engineering Standards Association committee.

R. E. STRADLING,
Director of Building Research.

BUILDING RESEARCH STATION,
Bucknall's Lane, Garston,
Watford, Herts.

January, 1927.

CONTENTS

LIST OF FIGURES

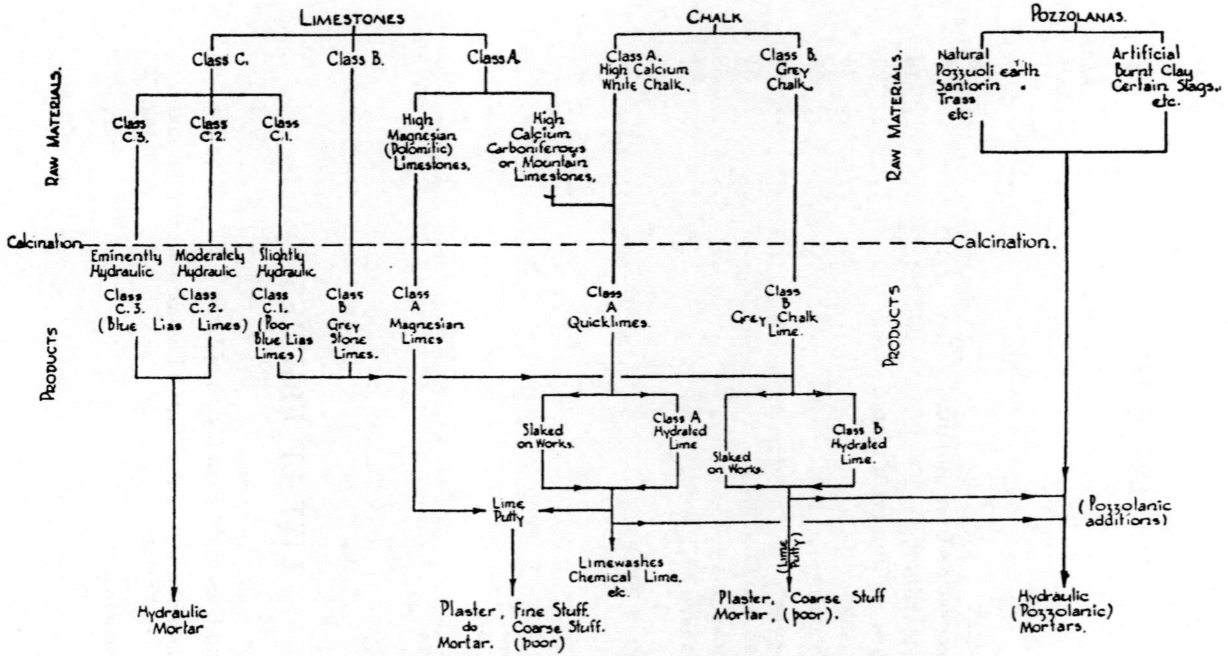

FIG. 1.—Chart showing uses of different types of lime.

Note.—This chart is intended only to give a general idea of the uses for which the different broad types of lime are best adapted, and should not be interpreted too literally. In particular, the distinction between Class B and poor Class C 1 limes is not clearly defined in practice; again, the oolitic limes are variable, and cannot be assigned bodily to the A Class, though often of the "High Calcium" type; whilst other limes than the Blue Lias may fall in the C Class. Pozzolanic additions can also be made with advantage to Class C 2 and Class C 3 limes.

Marls, marble, coral, and other forms of calcium carbonate which are not used to any great extent in the British Isles as sources of building lime, have been omitted from the chart.

LIME AND LIME MORTARS

INTRODUCTION

THE purpose of this Report is to give an account of the various uses of lime (quicklime and slaked or hydrated lime) in building operations, to give a practical classification of the different types of limes, and to indicate for which uses they are best adapted ; to collect together the many valuable hints and suggestions offered by traditional practices in the craft, in which the literature of the subject abounds ; and to provide a useful bibliography of references to accessible works for those who are sufficiently interested in particular points to wish to follow them up further. Detailed definitions and specifications of different lime products (and immediately related materials used in lime mortar and plaster) are then developed from the considerations brought forward here, and with due reference to such tentative or standard specifications as have already been prepared by different public bodies for lime products, gypsum plasters, and Portland cement.

It has been necessary to exclude more than a passing reference to certain materials used in building operations in which lime is but a relatively minor ingredient, and in the Section on Plaster to ignore the several excellent modern substitutes for straight lime plaster now available to the builder, since these do not fall in the category of quicklime or hydrated lime products. Similarily, the manufacturing processes and mechanical equipment of the lime-kiln are not dealt with here.

It might be suggested that a whole Section could, with advantage, be devoted to the subject of " Hydrated Lime," the fine powdered mechanically hydrated product which has become so prominent in late years in other countries, and which is rapidly gaining ground in Great Britain. However, this is dealt with in its various aspects in the course of the Report.

In the first Section, a brief account is given of the development of the uses of lime in historic times, and attention is drawn to certain ancient practices from which, even at this time, we can learn a great deal, even though these practices were quite empirical and often based on totally false theoretical notions. Suggestive recent developments are also outlined. In Section II a necessarily brief account is given of the actual processes and chemical actions involved in lime-burning, emphasizing the differences noticed between the behaviour of the various classes of lime at this stage. In the third Section alternative modes of classifying limes are described, and a regular scheme of classification into five classes—A, B, C 1, C 2, C 3—is given in detail; together with simple tests, based on Vicat's work, to distinguish practically between these classes. An account is also given of the geological distribution of the different classes of limestones in Great Britain. Section IV indicates briefly the ground to be covered in the subsequent Sections on individual uses of lime in building; Section V gives the results of recent work to determine the most reliable and lasting limewashes and cold-water paints with lime as a basis, suitable for making up on a small scale; and Section VI deals with the subject of lime plaster, with a discussion of common faults in plaster and their practical remedies, together with a consideration of plasticity, and its practical measurement. A brief account is also given in Section VI, for those who do not happen to be familiar with this branch of building, of different types of ordinary utilitarian lime plaster-work. Section VII deals with the important subject of pozzolanic additions to mortars (and cements), and gives a simple practical test for pozzolanic properties that can be applied to materials in use; Section VIII deals in some detail with the subject of lime mortars, with a reference to lime concrete; the paramount importance of using either a good hydraulic (Class C) lime, or else an approved pozzolana with Classes A and B limes, in good-class work, is explained fully here. Section IX deals with formal definitions and suggested specifications; the last Section, X, gives a brief review of the more readily available books of reference which cover the ground, at least in part; numerous references of a special nature, dealing with individual problems, are given at the end of each Section. In many cases, naturally, the same work will appear in both categories.

Throughout the Report, an attempt has been made to retain reasonable brevity and scientific accuracy. A complete understanding of such matters as the chemical actions involved in the burning of an eminently hydraulic limestone, or of the geological classification, is not essential for the understanding of the main principles and practical conclusions dealt with here.

SECTION I.—HISTORICAL

LIME and lime products have been in use from a very early period in the history of civilization. Amongst lingering remnants of primitive tribes which still exist at the level of stone-age man, examples are sometimes found of a utilization of burnt lime ; and various types of primitive kilns are used for making it. The phenomena involved in the dissociation of calcium carbonate by heat—lime-burning—must have been discovered soon after fire, by the use of limestone for hearths or boiling-stones, &c.

Early Traces

The earliest traces of the use of lime in constructional work are of lime plaster and stucco in early Egyptian buildings, chiefly of the monumental type. Prof. Sir Flinders Petrie has brought to light interesting examples of Egyptian plaster in pyramids and temples. Some of those in the chambers of the pyramids, of date around 2000 B.C., are still hard and intact. Part of the work was apparently executed in a mixture of lime and burnt gypsum plaster —the finest is stated to be in gypsum. Three-coat lime stucco, ¾-in. thick, on reed laths and with a haired gypsum finish has also been observed in these ancient buildings. A lime-gypsum plaster was used at times in the masonry of the pyramids, to fill joints and level up hollows ; this was also used over sculpture, to provide a surface for colouring. An ancient hand-float, exactly resembling a modern plasterer's implement, is in Prof. Sir Flinders Petrie's collections. " Fibrous plaster," similar to the modern material, was used for decorating coffins around 1500 B.C. ; and masks dating from 4400 B.C. have been discovered made of the same material. A good example of fine plaster-work is provided by a large pavement at Tel-el-Amarna, on brick, of 1-in. thick lime stucco with a trowelled finishing coat ⅛-in. thick, and highly ornate in design : this dates from 1400 B.C. and was apparently the floor of a part of the harem of King Amenhoted IV. Magnificent mural paintings on two-coat pure lime plaster on a clay backing, dating from 1500 B.C., have been discovered in the palace of Knossos in Crete.

The use of lime in mortar in brick or stone construction is rarely observed amongst these ancient peoples ; mud, bitumen, and (particularly in Asia Minor) a crude burnt gypsum were the materials most frequently in use, though indeed many large buildings which survive, in part, to this day were erected without any mortar at all between the heavy dressed stones.

The ancient Greeks made much use of lime stucco on their temples. In Homer's time temples at Mycenae were covered with a fine lime stucco ; about 500 B.C. this was the common external covering to temple buildings, even when built of marble. Stained plaster pavements were common. Pliny mentioned that in 450 B.C. a polished lime stucco mixed with milk and saffron was applied in a temple at Elis. In Rome there was much plaster-work, this being the common type of external finish for private houses; traces of these were mainly destroyed in the fire of Nero, A.D. 64, and in rebuilding. A few exquisite fragments of their decorative modelled stucco are still preserved in Rome; and some also in Pompeii (therefore dating before A.D. 79).

In illustration of the care taken with their materials, one may quote Pliny the Younger, who stated that "no builder should employ lime which had not been slaked at least three years," and mentioned that the Greeks used to grind their lime very fine and beat it with pestles of wood.

The greatest classical authority on architectural subjects, Vitruvius, who held a position analogous to that of Director of Ordnance to one of the Caesars and had had experience as a practical architect as well, gives most explicit working instructions for the choice and use of lime, sand, pozzolana and other materials for mortar and plaster; and full details for the application of fine six-coat marble-dust and lime stucco on reed laths for internal decoration work. This stucco was worked and rubbed until it became extremely hard and polished; in fact, portions of old stucco were sometimes treasured for use as mirrors.

In later classical times the Romans were noted for their fine public works executed in stone and hydraulic lime mortar, the mortar being made with an addition of the pozzolanic sands abounding in the volcanic districts around Baiae and Mount Vesuvius. The addition of ground potsherds (giving a similar effect to that of pozzolana) was also commonly recommended. The great Roman aqueducts were erected and waterproofed with the aid of such hydraulic mortars, obtained from sources which have remained in use to this day.

Mediæval Times

In the Dark Ages following the decline of Rome's power, in common with all other branches of art and learning, the builder's craft became debased, and the art of making hydraulic mortars was lost: in mediæval times, castles and cathedrals were erected with soft lime mortar, and such as survive owe their strength to massive design or to the cunning of the designer rather than to any great strength of the mortar, which is generally found to be in a soft and crumbly condition in the interior of walls and piers. In plastering, mud would appear to have been more common than lime, though the Anglo-Saxons plastered their houses inside and out, and the sanitary and fireproof qualities of lime plaster were appreciated. King John, after a disastrous fire had consumed London Bridge, issued an edict in A.D. 1212 to the effect that "all shops on the Thames should be plastered and whitewashed within and without—within eight days." In the 13th century lime was sold by the bag and cwt., and was used mixed with sand or pounded tiles (the latter giving a pozzolanic effect). Many mediæval houses and churches were plastered; and a rough-cast of lime, hay, coarse sand and pebbles was in use.

Renaissance

With the revival of interest in classical learning and classical arts during the Italian Renaissance an intense study was made of the ancient arts of decorative plaster and stucco work, and in the hands of Raphael and others an extremely high development of elaborate stucco was rapidly attained in Italy, which after a period of great activity ultimately reached France and England. The movement in England reached, perhaps, its climax in the famous "Nonsuch" royal palace, near Sutton, about 1600. Contemporary writers describe this large palace, now entirely vanished, as being covered externally and internally with masses of artistic figure-work in stucco, illustrating legends, &c. Possibly this work was executed in a plaster made by mixing rye dough with the lime, the gluten of the grain retarding the setting to some extent. Gypsum plaster came into use in France about this period.

A contemporary recipé for a fine lime plaster shows the influence of Vitruvius, who was accepted then as a supreme authority: "Take three parts of pounded Parian marble; add one part of lime which is to be perfectly slaked by letting it lie in a heap covered with pozzolana and exposed to the sun and rain for at least a year. Mix a day before with sufficient water on a tile floor. The first coat to be mixed with coarsely pounded, and the finishing coat with finely pounded white marble."

This great artistic development in the 16th century, of which perhaps one of the finest examples surviving is in the Holyrood Palace, slowed down later, and took an entirely different direction. Monotonous repetitive and cast work appeared, and plaster tended to become utilitarian rather than artistic. The Italian artist plasterers, the "Gentlemen Plasterers," entitled to wear ruffled shirts, lace, and to carry swords, disappeared; another school of architecture with other ideals became prominent.

Though similar materials were employed in both types of work, a distinction was drawn between plastering proper and "pargetry," the latter being purely trowel work done in places where a float and rule could not be conveniently applied, as in timber-framed houses. The last term survives in the "pargetting" or "parging" of chimney-flues. Some of the pargetry was of an elaborately ornamental type, designs and patterns being impressed in the soft plaster by means of stamps; a few examples survive to this day. The plasterers and pargetters were formed into a common Guild and Company in 1501 by a charter granted by Henry VII. The growth of specialization in the building arts is illustrated by an Act of James I forbidding painting by plasterers in the City of London; and by the Charter of Charles II which forbade any person to carry on simultaneously the trades of mason, bricklayer and plasterer, and required seven years' apprenticeship in the last-named craft.

Subsequent history shows the gradual encroachment of plasters and compositions, mainly based on burnt gypsum, on the sphere of lime, and the development of fibrous plaster and Portland cement stucco to replace the straight lime plasters, cement stucco in particular for exposed and external work.

Only a passing reference can be made here to the high development of plaster work in Asia Minor, and amongst the Moors, since the Persian and Saracen work depended mainly on gypsum plaster.

The fine plaster-work of India depended for its finish and durability upon patient labour and repeated operations in which time was of no account, and on the thorough working and beating of the material on the wall with wooden mallets whilst setting took place.

The ancients laid great stress on beating the plaster before use, for tempering this when used in fine plastering; it has been suggested that modern lime plaster has less durability than the ancient examples as a result of omitting this treatment. A general tendency towards this omission appears to be contemporary with a significant and historic strike of the "hawkboys" or plasterers' assistants in the London area. Part of the hawkboys' duty was the beating of the mixed plaster before application.

Mortars

Until John Smeaton made his painstaking investigations in 1756 in connexion with the building of the Eddystone lighthouse, and as a consequence revived the almost forgotten technique of the preparation of a true *hydraulic* mortar, which does not rely mainly upon a fortuitous re-carbonation to develop its strength, little progress had been made in the art of mortar-making; the presence of gypsum and of soft mortar still containing up to 20 per cent. of free (uncarbonated and uncombined) lime in the piers of St. Paul's Cathedral illustrates this. However, traditional Indian building methods show an appreciation of pozzolanic additions, in the form of "surkhi" or burnt clay; and a hydraulic mortar has long been in use in India, made from an argillaceous limestone "kankar." Hydraulic mortars made with the Italian pozzolana, and the German trass of similar properties, have also had a limited local use for centuries.

In 1756 Smeaton, though handicapped by a lack of any extensive knowledge of chemistry, set himself the task of discovering what was the particular characteristic of British hydraulic limes; and in what way the necessary quick-setting properties of such a lime, when immersed wholly in water, might be developed. He found that a Blue Lias lime from Aberthaw, containing a proportion of insoluble clay, (when properly slaked and worked up) showed the best hydraulic properties; and he imported the classical Italian pozzolanic earth from Civita Vecchia, near Rome, to mix with this, thus preparing a genuine hydraulic cement of great strength.

The Invention of Portland Cement

Smeaton's work was followed by the patenting, in 1824, of Aspdin's Portland cement. *Natural* hydraulic cements, sometimes called " Roman Cements," made by burning argillaceous limestone nodules found in the London clay and in the shale beds of the lias formation, and other limestone deposits unusually rich in clay, came into use, particularly where their extreme rapidity of setting gave them an important advantage. But the uniformity and regularity of supply possible with artificial Portland cement manufactured from lime and clay in the right proportions has latterly brought about an almost complete supersession of natural hydraulic cements by the manufactured article, except where large deposits of a uniform natural cement-stone are available, as in New York State, and the Louisville and Lehigh districts of the United States, where a number of natural cement plants are still in operation. Even in the U.S.A., though, there was a large decrease in the period 1903–13 in the use of pozzolanic and natural cements, and but a small increase in the use of building lime in the same (pre-war) period. (Eckel.)

Additions to Lime Mortar

A curious aspect of the history of the use of lime is afforded by the array of materials which have been added, from time to time, to lime plaster and mortar, generally with a view to slowing down the setting and making possible a more extended manipulation of the material in artistic work. Thus fig and other fruit juice, elm bark (presumably in the form of an infusion), barley-water, bullock's blood, cow-dung, hot wax, white of egg, wort and beer, pitch, milk, gluten, butter-milk, cheese, saponified beeswax, &c., have been recommended. White of eggs and wort were used with the lime in the old Charing Cross in London; sugar (" jaghary ") is apparently a common addition to lime mortar in India; and the modern use of glue or size-water in the gauging of plaster is familiar. Whilst no definite accelerator for hastening the setting of ordinary lime mortar has become prominent (beyond the normal practice of gauging with Portland cement or burnt gypsum for this purpose), recent patent specifications* as well as a suggestive pronouncement of the United States Bureau of Standards appear to suggest immediate possibilities in this direction.

Recent Developments

An interesting recent development is in the direction of the use of high magnesian limes for plaster and for general building purposes, and even for hydraulic cements, particularly in the

* U.S. Patents Nos. 1554183 and 1554184 of 1925: Holmes and Fink, *vide* Section VI.

United States and Italy. A revival, under official encouragement, of the use of the German pozzolanic material, the *Trass* which is found particularly in the neighbourhood of Andernach, on the Rhine, is to be noted; as well as recent experiments in the use of other pozzolanic additions, such as spent oil-shale, granulated slag, Japanese volcanic ash, &c.

Perhaps the most important recent development has been the marketing of mechanically hydrated or slaked lime, of uniform quality and in extremely fine powder, for preparing lime putty or mortar directly by the addition of the necessary water. This movement has made great headway in the United States.

Another interesting development is the sale, in large cities, of ready-mixed mortars, the material being delivered ready for use in tank wagons. Ready-mixed dry plaster, requiring only the addition of water for use, has already a considerable vogue in some parts.

References

BANKART, G. P., *The Art of the Plasterer*. London, 1908.

BLOUNT, B., *Cement*, p. 4 *et seq*. London, 1920.

BRIGGS, M. S., *A Short History of the Building Crafts*. 1925. Oxford University Press.

BURCHARTZ, H., *Mitt. Materialprüfungsamt*. 1912, 1921, etc. (Esp. on Trass.)

DAMM, H., *Kalkstein und Kalk bei Naturvölken und im Altertum*. Berlin, 1922.

DESCH, C. H., *Chemistry and Testing of Cements* (especially p. 10, lime concrete). London, 1911.

DIBDIN, W. J., *Lime, Mortar and Cement* (especially p. 147, Chinese concrete). London, 1901.

ECKEL, E. C., *Cements, Limes and Plasters*. London and New York, 1922.

FLINDERS PETRIE, *Arts and Crafts of Ancient Egypt*. Edinburgh and London, 1909.

HASAK, *Tonind. Ztg*., 1925, **98**, 1385.

KNIBBS, N. V. S., *Lime and Magnesia*. London, 1924.

MILLAR, W., *Plastering, Plain and Decorative*. London, 1897.

PLINY, *Hist. Nat. Bk*. **36**, Chapters 52, 54, 55.

QUIETMEYER, *Zur Geschichte der Erfindung des Portlandzementes*. Berlin, 1912.

—— *Roorkee Papers*. Thomason Civil Engineering College, India, 1900, **1**, Part I.

—— *Roy. Inst. Brit. Arch. J*., 1922 (Aug. 19), 1925 (July 18), 581.

SIMPSON, *History of Architectural Development*, 1905, **1**.

—— *Soc. of Mural Decorators and Painters in Tempera, Papers*, 1925, **2**.

VITRUVIUS, *Architecture*, Books II and VII. Translated by Gwilt. London, 1880.

SECTION II.—THE PROCESS OF LIME BURNING

The Chemical Process

THE chemical process involved in lime burning consists essentially in the dissociation by heat of calcium and magnesium carbonates; the removal of the carbon dioxide resulting from this action; and the subsequent cooling of the oxides of calcium and magnesium, which are the other products of the reaction, in an atmosphere fairly free from carbon dioxide, a precaution necessary in order to avoid recombination.

Types of Lime-Kilns

Only a brief account of the technical processes of lime manufacture will be given here. While a large number of different varieties of lime-kilns are in use throughout the world, the essential process is the same in all, and they may all be divided into two general types, the *Intermittent* and the *Continuous*. The primitive heap of coral and driftwood of the South Sea Islander, and the crude shaft in the side of a chalk pit common in some parts of England ("flare kiln"), are examples of the first type; whilst the continuous kiln has been highly developed into vertical kilns with mixed feed; vertical kilns with separate feed (including producer-gas and natural-gas-fired kilns); rotary kilns; and ring or chamber kilns.

The intermittent kiln in a more developed form consists of a wide chimney or vertical shaft, which may or may not be lined with fire-brick, open at the top and with a side opening at the bottom. A fire is built at the foot, under a rough arch of limestone lumps, and the rest of the charge is heaped up above the arch. After several days the fire is allowed to burn out and the furnace to cool, and the burnt lime is removed through the side opening.

In vertical continuous kilns with mixed feed ("pot kilns"), limestone and fuel are charged in alternately at the top of a similar wide chimney or shaft, and the burnt lime is removed intermittently at the bottom, mixed, of course, with the ash from the fuel used. In the separate-feed continuous kilns ("patent kilns"), only the limestone is charged directly into the kilns, the fuel being burnt (if in the form of coal or wood) in fireboxes at the sides which open into the kiln at points near the lower end; with gas-fired kilns the gas is injected at a corresponding point and burnt in contact with the charge; in these the ash of the fuel is kept almost completely separate from the burnt lime. The rotary kiln resembles exactly the rotary Portland cement kiln, a fire-brick-lined cylinder being slowly rotated in a position slightly inclined to the horizontal, whilst raw limestone (in small pieces) is charged into the upper end (which is also connected to a chimney), and the burnt lime withdrawn continuously at the lower end; the heating being carried out, as in the cement kiln, by a flame from an oil-burner or powdered coal-dust injected by compressed air. The ring or chamber kiln, used largely in brick-making, and to some extent also in Portland cement manufacture, has a ring of chambers communicating with one another and with a central chimney-stack by flues, which can be closed by dampers. By proper manipulation of the dampers, each chamber in turn becomes fireplace, burning zone, pre-heating zone, drying zone, and chimney connection. In both rotary and ring kilns, the fuel ash is left in the finished product. The economics of the different types will not be discussed here.

The Chemistry of Lime Burning

Chemically speaking, the raw material (limestone, chalk, marble, coral, &c.) must contain as its essential constituent calcium carbonate ($CaCO_3$), and may contain varying amounts of magnesium carbonate ($MgCO_3$), the latter ranging up to 43 per cent. of the whole; in a high calcium lime (such as is commonly obtained in England from the chalk, oolitic, or mountain limestones) the magnesium represents only a small proportion of the metallic bases present, calcium predominating. However, in some countries high magnesium limestones, approaching dolomites in composition, are burnt for the production of building lime, and give good results; it is more than likely that high magnesium limes have been in use at many times and places without the users being aware that they were not handling a nearly pure calcium lime. Accordingly, it is necessary to deal with the properties of magnesium limes in conjunction with those of approximately pure calcium limes.

Hydraulic Limes

As indicated here, the inevitable impurities (*i.e.*, that portion of the limestone which is not calcium or magnesium carbonate, or useful clayey matter) present in the raw material should not exceed 5 per cent. by weight. As will be shown in Section III, "Classification," of this Report, there is present in *hydraulic* limes a very valuable ingredient (not an "impurity") in the form of argillaceous matter, *i.e.*, *Clay*, consisting, chemically, mainly of a hydrated silicate of aluminium.

Accordingly, in a good *hydraulic* limestone :—

<table>
<tr><td rowspan="2">in addition to</td><td>calcium carbonate</td><td>=</td><td>lime (CaO)
plus
carbon dioxide (CO_2)</td></tr>
<tr><td>and magnesium carbonate</td><td>=</td><td>magnesia (MgO)
plus
carbon dioxide</td></tr>
</table>

that are present in ordinary non-hydraulic limestone, there will be :—

$$\text{Clay matter} = \text{hydrated aluminium silicate} = \begin{cases} \text{alumina } (Al_2O_3) \\ \text{plus} \\ \text{silica } (SiO_2) \\ \text{plus} \\ \text{water.} \end{cases}$$

It is the nature of this clayey matter and the ratio it bears to the lime that determines whether the lime be " fat " or " hydraulic."

Generally, there will be, in addition, greater or smaller traces of iron ; of alkali metals (sodium and potassium) ; and of sulphur, either in combination with iron as sulphide (*i.e.*, iron pyrites) or in combination with some of the bases present as sulphates (*i.e.*, as gypsum, Epsom salts, or Glauber's salts). A varying amount of " *insoluble residue* " may also be reported in an analysis, probably consisting mainly of silica sand (quartz) insoluble in ordinary dilute acids and distinct from clay, which contains *combined* silica ; this residue should not exceed 5 per cent. of the total weight.

The chemistry of the burning process is greatly complicated by the presence and interactions of the argillaceous ingredients, the chemical reactions involved approaching in complexity those found in the burning of Portland cement.

Burning of Non-Hydraulic Limes

With a non-hydraulic or but feebly hydraulic lime the chemical action in burning is simple. As the temperature of the stone is gradually raised in the lime-kiln, first the free water is driven off in the form of steam, and passes out of the kiln together with the products of combustion of the fuel used and superfluous air drawn through by the draught. Then as the temperature approaches a white heat (900–1200° C.), the carbonates begin to decompose and give up their carbon dioxide ; at first from the surface of the lumps, as these are heated up first, and then from the interior, the carbon dioxide diffusing out through the pores. If the product, the *carbon dioxide*, were not removed from the sphere of action, the dissociation of the limestone would soon come to a standstill, *i.e.*, as soon as a certain pressure, the dissociation-pressure at that particular temperature, had been attained.

Use of Steam

It has been suggested, and it is actually the practice in some lime-works, to increase the dilution of the kiln gases by increasing the amount of steam, in order to prevent any recarbonation of the lime. To effect this, either wet materials are used or steam is actually injected. There does not appear to be any very conclusive evidence for or against the practice.

Continuous Kilns

In vertical continuously-operated kilns the preliminary drying out of most of the moisture in the stone is carried out in the upper part of the kiln, which in more elaborate installations is often arranged to hold a large charge, sufficient to last for many hours. Then as the material descends in the kiln, it approaches the hot combustion-zone, and attains a temperature high enough to determine the thermal decomposition of the calcium and magnesium carbonates. For calcium carbonate this is about 900° C.; for magnesium carbonate 550° C. The white-hot porous stone is then cooled either in the lowest part of the kiln, where it meets a current of air, or by being passed at once into the open air. The heated air in the first case is used for supplying the kiln, thereby effecting an economy of fuel. In either case, the air which comes into contact with the hot lime is sufficiently free from carbon dioxide to prevent appreciable recombination.

Nature of Product

The chemical composition of the product depends again on the nature of the stone used; ranging from nearly 100 per cent. calcium oxide for a pure high-calcium stone to 58 per cent. for a good dolomite, the balance being mainly magnesium oxide in the latter case. There should be only a trace of carbon dioxide present in fresh burnt lime and, of course, no moisture. In mixed-feed kilns there is also present the ash of the fuel.

The physical condition of the product depends on the stone burnt. In general, it will be porous and friable; some stones break up readily in the kiln and will produce a large proportion of fine material and dust. Irregular burning will produce an excess of underburnt lumps, containing a core of carbonate; overburning, produced either by a generally excessively high temperature or by a too prolonged stay in a normally heated zone of a certain proportion of the stones, gives hard, often yellowish, lumps, which take an extreme time to slake and are, in general, useless. Experiments made with specially pure limes appear to indicate that the temperature at which overburning occurs is lowered by the presence of impurities; chemically pure lime cannot be overburnt at any temperature usually possible in a kiln. It follows that magnesian and hydraulic limes are more easily overburnt than high calcium limes. The effect is probably due to sintering or partial fusion of the material, causing the formation of crystalline lime particles of appreciable size; whilst when properly burnt the lime is in such extremely fine particles that it loses its crystalline properties.

Temperatures

The actual temperature of dissociation of calcium carbonate has been determined by a number of observers, the figures obtained differing slightly; a recent determination by Andrussow of the point at which the dissociation pressure is equal to the mean atmospheric pressure, *i.e.*, the minimum effective temperature for burning lime, gave 882° C. Former determinations by Johnston, 898°, and Mitchell, 896° C., are in fair agreement with this. Other investigators have given figures up to 910° or even 916° C. For pure magnesium carbonate the figure is about 550° C.

If suitable precautions are taken so that the temperature remains throughout below 800° C. and the partial pressure of carbon dioxide in the kiln is maintained at a higher figure than is otherwise usual, all the magnesium carbonate may be calcined without decomposing any calcium carbonate. This method has been suggested as a means of utilising impure magnesium carbonates for making magnesium oxychloride cements, where the presence of 2 per cent. of calcium oxide is undesirable.

The actual temperature in the burning zone in a normal kiln will be considerably higher than 882° C., since the outside of a lump must be much hotter in order that the centre portion may reach this temperature in a reasonable period, so that as high a temperature as possible (without overburning an appreciable proportion of the stone) is desirable in order to save time : from 1000° to as high as 1400° C. has been observed. An ordinary figure is 1200° C. (= 2192° F.). A certain definite amount of heat must be transferred to the stone to effect the decomposition and liberation of the carbon dioxide, after the decomposition-temperature has been reached ; this is approximately 770 B.Th.U. per pound (43,300 cal. per gm. mol.) for pure calcium carbonate and 464 units for pure magnesium carbonate. So that in the lime-kiln, heat is required and fuel is consumed, first, to heat the stone to the boiling point of water and to evaporate free moisture ; second, to heat the dry stone up to the dissociation point ; third, to provide for the heat of dissociation ; and fourth, to heat up the air and any additional steam provided, which pass through the kiln. In some kilns part of the heat is saved by passing the incoming air over the hot burnt lime, thus both hastening the cooling of the lime and preheating the air. In the Eldred process some of the kiln gases issuing from the kiln top are returned by a blower under the grates, together with fresh air to provide the necessary oxygen to support combustion.

Fuel

As a long flame of but moderate temperature is desirable in lime-burning, rather than an intensely hot short zone of combustion, wood provides a very favourable fuel in a lime-kiln ; and when coal or gas, &c., are used from motives of economy an attempt is made as far as possible to imitate the conditions in a wood-fired kiln. It is evident that in a mixed-feed kiln, where the fuel is charged in together with the limestone, the ash of the fuel will remain to contaminate the product. Since in some large kilns the weight of the fuel is around one-third of the weight of the lime produced, the composition of the ash is a matter of considerable importance when a high degree of purity is desired, as is the case for many chemical purposes ; in such cases either gas or externally fired kilns are preferable.

Screening

Except to satisfy a popular prejudice for a lime almost exclusively in large lumps, there is no benefit in separating (either by screening or forking) the lump from the fine and powdery portions of the product of externally fired kilns. Actually the large lumps are more likely to be overburnt outside or to contain a core of underburnt material than the powder, though the latter is, at times, viewed with suspicion by users on account of its superficial resemblance to the powder produced by air-slaking of lump lime, which powder consists of hydrated and partly carbonated material. When non-hydraulic limes are hydrated immediately for marketing as hydrated lime, a recent development, no preliminary screening is required, though before entering the hydrators the largest lumps may be crushed to a more convenient size. After the hydration, however, the lime is screened or air-separated.

Burning of Hydraulic Limes

As indicated above, the chemical actions are more varied and complex during the formation of hydraulic limes than when practically pure carbonates are being dissociated by heat. In addition to the dissociation of the carbonates present, all the hydrated compounds present as clay material (hydrated aluminium or iron silicates) are first decomposed ; up to 900° pozzolanic compounds are formed, and above this temperature begin to decompose ; then above 1000° C. calcium silicates and aluminates begin to be formed, most of which are eminently hydraulic in nature and are the cause of the cementing properties of Portland cement. In most cases the silica will be largely in excess of the alumina ; a burnt hydraulic lime then consists essentially of a *mixture* of ordinary quicklime with a calcium silicate, and on addition of water will slake, though slowly. Subsequent hardening depends mainly on the further hydration of the silicate, now reduced to a fine powder.

If the hydraulic lime is burnt at a sufficiently high temperature, some of it may sinter and produce clinkers (" grappiers ") which, like Portland cement clinker, are relatively inactive until very finely ground ; then from these grappiers a natural cement may sometimes be obtained rivalling the artificial Portland in strength. It is thus seen that the properties of the product of a burnt limestone may vary greatly, depending on the rate and extent of the heating.

References

ANDRUSSOW, *Zeits. phys. Chem.*, 1925, **116**, 81.

BLOCK, B., *Das Kalkbrennen.* Leipzig, 1924.

ECKEL, E. C., *Cements, Limes and Plasters.* Loc. cit. Chapters VII, XIII–XVI.

JOHNSTON, *Amer. Chem. Soc. J.*, 1910, **32**, 938.

KEPPELER, G., *Z. angewandte Chem.*, 1925, 397.

KNIBBS, N. V. S., *Lime and Magnesia*, Chapter VII, loc. cit.

LE JEUNE, EMILE, *Guide du Chaufournier*. Paris, 1900.

MITCHELL, *Chem. Soc. J.*, 1923, **123**, 1055.

MITCHELL, C. F., *Building Construction, Advanced Course*, Chapter I. 6th edition. London, 1909.

RIVINGTON, *Notes on Building Construction*, Part III, Chapter III, p. 193. 7th edition, new impression. London, 1919.

SMYTH and ADAMS, *Amer. Chem. Soc. J.*, 1923, **45**, 1167.

U.S. BUREAU OF STANDARDS, *The Manufacture of Lime*. Technologic Paper No. 16, 1913.

U.S. BUREAU OF MINES, *Manufacture of lime from small stone with a sintering machine*. Reports of Investigations. W. M. Meyers. Serial No. 2762. July, 1926.

SECTION III.—CLASSIFICATION

Methods

As indicated in Section II, limes are composed of calcium oxide together with varying amounts of magnesium oxide, depending on the nature of the stone from which they were burnt ; and in addition will contain varying amounts of argillaceous matter (clay), which, since it is the cause of valuable hydraulic properties, is not necessarily an undesirable impurity ; other compounds are nearly always present which are impurities of no value, and the latter should not exceed 5 per cent. by weight of the total.

The classification of limes is usually founded on the proportions of these ingredients present ; and upon the resulting behaviour of the burnt lime when brought into contact with water. A further subsidiary classification is used to denote the physical condition of the lime, whether in lump or in powder, or as to whether it is " Selected " *i.e.*, a well-burnt lime picked free from ashes, core, clinker and other foreign material, or " run of kiln," *i.e.*, without selection (American Society for Testing Materials Classification). A special designation, such as " chemical lime," may refer to exceptional chemical purity for certain manufacturing processes in which this is essential.

A distinction can be drawn between :—Calcium limes, containing 85 per cent. or over of calcium oxide ; and magnesian limes, containing over 10 to 20 per cent. of magnesium oxide. Some Italian limes will contain up to 40 per cent. of magnesia, and 4 to 9 per cent. is common in " fat " Italian limes. Eckel states that while an American lime containing over 5 per cent. of magnesia may be classified as " magnesian," such limes will usually contain over 30 per cent. of magnesia. American classification distinguishes high calcium, calcium, magnesian, and high magnesian limes respectively (American Society for Testing Materials ; and American National Lime Association).

From a chemical standpoint, the principal point of distinction is the type of *impurity* present. However, the estimation of the hydraulic properties of clay substance is a matter of considerable difficulty, especially when, as is the case at the moment, the particular compounds to which these properties are due are not known with any certainty. Undoubtedly they are due in many cases to a particular type of active silica often known as " soluble silica "* but this is not necessarily true in all cases. Hence we are, at present, limited almost entirely to practical tests.

* A method of estimating this " soluble silica " is in process of development in the Building Research Station.

CLASSIFICATION SCHEMES—VICAT'S CLASSIFICATION

	Class				
	Fat	Lean	Hydraulic		
			Feebly	Moderately	Eminently
	A	B	C 1	C 2	C 3
Slaking	Rapid : much heat	Moderately slow (5 min.)	Slow (5 to 60 min.)	Very slow	Very slow ; requires fine grinding.
Expansion on slaking	2–3 times	Large	Small	Small	Small.
Set in water	——	——	> 20 days	15–20 days ..	2–4 days.
Consistence after one year in water	Putty	Putty	←Intermediate→	Hard soap ..	Stone.
Residue on solution in much water	Very small ..	Small ; devoid of consistence	Appreciable ..	Insoluble	Insoluble.
Residue ditto in HCl	Ditto	Ditto	Ditto	Moderate	Large.
Composition (Vicat)	>94 % CaO MgO	>70 % CaO MgO residue inert	<12 % active clay, etc.	>12 % <18 % active clay, etc.	>18 % <25 % active clay, etc.

SUGGESTED CLASSIFICATION FOR BRITISH STANDARD SPECIFICATION

Suggested British classification $CaO + MgO$	>95 %	<95 % >85 %	>60 %	>60 %	>60 %
$SiO_2 + Al_2O_3$			<15 %	>15 % <25 %	>25 % <35 %
Tensile 1 : 3	—	—	—	> 75 lb.	3 days moist air. 25 days water.

CLASSIFICATION ADOPTED IN FOREIGN SPECIFICATIONS

German classification	" Air Lime " : >28 lb. Tensile 1 : 3. 24 hours moist air, 28 days air		>28 lb. tensile 21 days moist air 7 days water	" Strongly hydraulic " : >71 lb. 7 days moist air. 21 days water.	
Swiss classification	—	—	Hydraulic : >85 lb. Tensile 3 days moist air. 28 days water.		
French classification. Hydraulic modulus = % clay/ % lime	0 to 0·10		0·10 to 0·16	0·16 to 0·31 "Properly Termed " 0·31 to 0·42	0·42 to 0·50 " Limited " 0·50 to 0·65
			Set under water to resist finger in 5 days.		
American : U.S. Govt. Master Specification No. 250	Ca / Mg >95 % / Do. >75 % / — — / >20 % <10 % / Do. < 5 % / Do.	$CaO + MgO$ CaO MgO CO_2 $SiO_2 + Al_2O_3 + Fe_2O_3$			
South African (tentative) ..	Blue building magnesian lime <6·5 % SiO_2				
Russian	" Fat Air " ..	" Lean air " ..	" Feebly hydraulic "	" Strong hydraulic."	

The Vicat Classification

The most satisfactory classification is perhaps a slight modification of Vicat's historic scheme. In accordance with this, limes are divided into the following classes :—

Class A.—Rich or Fat Limes.
Class B.—Poor or Lean Limes.
Class C 1.—Slightly or Feebly Hydraulic Limes.
Class C 2.—Moderately Hydraulic Limes.
Class C 3.—Eminently Hydraulic Limes.

The differentiation of these classes is effected by the following simple tests as originally put forward by Vicat himself.

Testing Hydraulicity : Practical Tests

The lime, freshly burnt and broken to walnut size, is put into a cloth bag of open material or a basket, a volume of about a quart being used ; and is immersed for not more than five seconds in pure water, then drained a moment, and emptied into a stone or cast-iron mortar. The behaviour of the lime subsequently is closely observed: the lime may immediately commence to hiss, decrepitate, swell up and give off steam, and at the same time may fall into powder ; there may be a period of inactivity of 5 minutes, after which the same phenomena may be observed ; or only after 15 minutes have elapsed, in which case the decrepitation and steaming are much less marked ; a whole hour may elapse before any change is observed, when the action is again greatly moderated, so that but little heat is disengaged ; or in extreme cases there may be no visible change for long periods of time.

As soon as the lime shows signs of breaking up, water is added carefully down the side of the vessel, the lime is stirred up with a spatula, and finally is worked up by a pestle to the consistence of a stiff clay, more water being added if necessary. It is then left for a couple of hours, or until perfectly cold.

After standing, the lime paste is then worked up again with the pestle, with the addition of more water if required, to the consistence of potter's clay, and sufficient is placed in a small pot or glass jar, of greater height than width, so as to fill it approximately two-thirds full. The lime is caused to settle down in the pot by striking the bottom of it on the palm of the hand or on a block, and the whole is then placed standing in a vessel of water so that it is immersed. Any change in consistence of the material is then observed over a considerable time.

The division of limes into Vicat's classes follows then according to their behaviour during slaking and subsequent storing under water.

Class A or Fat Limes slake rapidly and energetically, and have their volume doubled, or more, in the process ; their consistence remains sensibly the same after many years of immersion in water under the conditions of the test ; and they dissolve practically completely in pure water frequently changed.

Class B or Lean Limes slake slowly, with little or no increase in volume, and behave in water in a similar manner to the Fat Limes, except that they leave a residue devoid of any consistence on solution in many changes of water.

Class C 2, Moderately Hydraulic Limes, set in 15 or 20 days after immersion, and continue to harden; but after six or eight months the process becomes slow. After one year their consistence is about equal to that of hard soap. They dissolve in water with great difficulty (leaving of course an appreciable residue); the expansion on slaking is small.

Class C 1, Feebly Hydraulic Limes, will be intermediate between this Class C 2 and Class B.

Class C 3, Eminently Hydraulic Limes, set at some time between the second and the fourth day of immersion. After one month they are already very hard and altogether insoluble. At six months they appear like the absorbent calcareous stones, splintering under a blow and presenting a slaty fracture. The expansion on slaking is small.

It will be seen that a certain amount of time is necessary even for the preliminary stages of these tests; accordingly the determination as to in which class a lime is to be placed cannot be done immediately, though a chemical analysis (by indicating the proportion of argillaceous matter present) will give valuable information to an experienced man.

As a guide, Vicat's approximate figures for the proportions of inert and argillaceous matter present in the different types of limes may be given here: Class A, below 6 per cent. of (inert) silica, alumina, iron, &c.; Class B, below 30 per cent. of inert impurities; Class C 1, below 12 per cent. of clay, &c. (as *active* ingredients); Class C 2, up to 18 per cent. of clay, &c., silica predominating; and Class C 3, up to 25 per cent. of the hydraulic ingredients, silica again predominating. It will be noted again that it is extremely difficult to distinguish, by analysis only, between silica and alumina which are merely inert adulterants of the lime, acting as so much fine sand; and silica and alumina which, as clay-substance, exert the favourable hydraulic effect. Modern practice allows of a higher percentage of hydraulic ingredients, up to 36 per cent. in a good eminently hydraulic lime.

Other Tests

To distinguish between a calcium and a magnesian lime a chemical analysis is both necessary and conclusive, though in the first instance suspicions may have been aroused by the relative speed and energy of slaking, and degree of expansion, exhibited in Vicat's slaking test, magnesian limes showing the least.

To distinguish a Class A lime, a small quantity of it is treated, after slaking in excess of water, with a little dilute hydrochloric or nitric acid. The amount of residue, estimated by eye, indicates the type: the rich lime dissolves at once and practically completely, the others will leave an appreciable quantity of fine residue devoid of structure or consistence. This test can be carried out in a few minutes in a cup or other small vessel. Fragments of the original limestone can be tested similarly: a high calcium or magnesium limestone will dissolve with effervesence, the magnesium with notably less energy, leaving practically no residue; a dolomite will dissolve slowly in cold dilute acid, but will finally go completely into solution if pure ; a hydraulic stone will leave an appreciable insoluble residue. A poor lean limestone will similarly leave an insoluble residue.

Hydraulic Index

The hydraulic quality of a lime may be defined more nearly by the value of the " hydraulic index," which is used in conjunction with cements. This is a ratio between the constituents of the lime, which is expressed thus :—

$$\frac{\text{Percentage of Silica} + \text{Percentage of Alumina}}{\text{Percentage of Lime}}$$

Ordinary limes show a hydraulic index less than 0·1 ; feebly hydraulic from 0·1 to 0·2 ; good hydraulic limes will give from 0·2 to 0·4, whilst Portland cements and natural cements run from about 0·4 to 1·5 and over ; a pozzolana will exhibit a hydraulic index over 3.

A more complex expression, the " cementation index," which distinguishes more precisely the effects of the silica, alumina, and iron oxide present, has been suggested by Eckel.

As the hydraulic index was devised in connexion with Portland cement, in which the magnesia content is always low, it does not include the latter at all ; whether a more suitable formula would result from its inclusion cannot be stated without further research.

Geological Distribution of Limestones

A possible classification of limes would be in accordance with the geological formations in which the stones occur. The larger masses of calcium carbonate are marine in origin, and were formed by the deposition of calcareous shells, &c., on the sea-bottom. These limestones, if occurring in areas free from terrestrial deposits, would be very pure, but where such deposits could take place more and more argillaceous matter (derived from the land) would be included. The pure limestone deposits would be a source of fat limes, whilst the others would yield either lean or hydraulic limes according to the nature of the additions. Further, the coast deposits would generally be more variable, both in occurrence and constitution. However, over such a relatively small area as the United Kingdom, the nature of the deposits would generally be of a similar type for a given geological formation; so that a geological classification gives at the same time a rough geographical classification, and is useful as an intimation as to what type of lime is likely to be most plentiful in a particular district, where a certain geological formation is commonly observed.

Lime-bearing Strata

Geological formations containing important amounts of lime-bearing rocks are to be found mainly in the Mesozoic (*e.g.*, Cretaceous and Jurassic rocks) and late Paleozoic (*e.g.*, Permian and Carboniferous limestones), and to a less important extent in the Devonian and Silurian subdivisions of the middle or older Paleozoic. The commonly used rocks are indicated in the accompanying table and also in the map showing the distribution of the chief limestone-bearing formations of Great Britain.

GEOLOGICAL TABLE

Period.	Strata.	British Limestones used in Lime Burning.
Quaternary	Recent	
	Pleistocene	
Tertiary	Pleiocene	
	Miocene	
	Oligocene	
	Eocene	
Mesozoic	Cretaceous ..	Upper Chalk: White Chalk. Middle Chalk. Lower Chalk: Grey Chalk; Chalk Marl (Clunch).
	Jurassic.. ..	Oolitic: (Purbeck, Portland, Bath and Inferior Oolite). Liassic: Blue Lias.
	Triassic.	
Paleozoic	Permian ..	Magnesian and Dolomitic.
	Carboniferous ..	Northern Limestones (thin-bedded). Southern Limestones (massive or "Mountain").
	Devonian (and Old Red Sandstone) ..	Devonian Limestone.
	Silurian ..	Grey Limestone.
	Ordovician	
	Cambrian	
	Pre-Cambrian	

The Chalks

The Cretaceous deposits of Great Britain, in the south-west and in Yorkshire, contain an upper layer of white chalk, producing a Class A lime. Below this is frequently a layer of grey chalk, containing some argillaceous matter, and yielding a lime corresponding either to Class B or C 1. Below this again occurs the chalk marl, "clunch,"* yielding a poor Class B lime.

The Blue Lias and Oolitic Limestones

The Jurassic strata possess limestones which are important building stones (*e.g.*, the well-known Purbeck, Portland, and Bath stones); when burnt for lime these Oolitic (resembling fish-roe in

* This term is also applied as a generic term to all chalk rocks that can be utilized for building purposes. *Vide* J. Watson, British and Foreign Building Stones, Cambridge, 1911, p. 197.

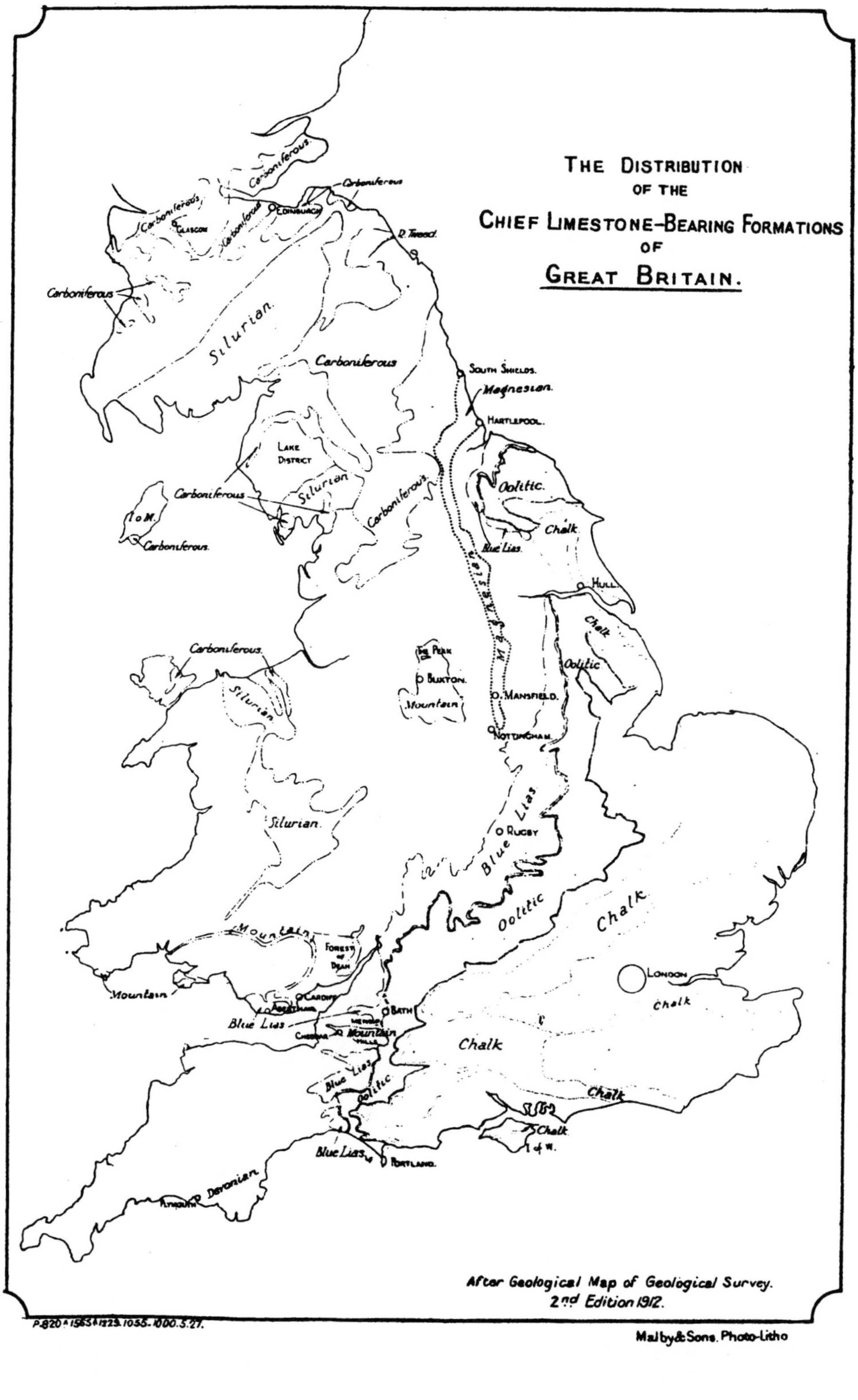

P.820 A 1565 & 1223. 1055. 1000. 5.27.

Malby & Sons. Photo-Litho

appearance) limestones may give a Class A lime. The Liassic includes the famous Blue Lias limestone, containing sufficient argillaceous matter to yield a good hydraulic Class C 2 or C 3 lime.

The Jurassic rocks outcrop in a comparatively narrow band extending continuously from Portland to Hull, with an extension on the north coast of Yorkshire ; the Lias forming mainly the more westerly portion of this belt, with an extension to S. Wales and Somerset. There is considerable local variation in the Blue Lias, the best (*i.e.*, C 3, Eminently Hydraulic) being associated with definite districts, such as Aberthaw, Rugby, and Barrow-on-Soar. Others may fall rather in the Class B (Poor or Lean), and may even approach a calcareous shale in composition ; the colour of the lime is not necessarily a guide to its hydraulic properties. Such local variation is common in the case of hydraulic limes, their properties depending on the nature of the clay present and this, as has been shown, depending on their relative position during the original deposition of the limestone.

Magnesian and Dolomitic Limestones : Permian

The limestones in the Permian division are characteristically rich in magnesium, ranging from occasional examples of high calcium limestone through magnesian limestones to a pure dolomite containing approximately the theoretical proportion of 45·65 per cent. of magnesium carbonate. These form a narrow belt, parallel to and west of the Jurassic, running from Nottingham to South Shields, and widening out at the top to reach Hartlepool. They do not outcrop south of Nottingham.

Carboniferous Limestones

The most important of the carboniferous limestone series, often referred to as " mountain limestone " when in massive deposits, occur below the coal measures ; these frequently give on burning a Class A lime. They are found in the Mendip Hills and at points in Somerset, surrounding the Forest of Dean and South Wales coalfields, at points in North Wales, and in a wide field in the Peak district. Carboniferous limestone also occurs in the north of England and the south of Scotland ; but here it has nearly always lost its mountain character, and has become divided into thin beds separated by shales, grits, and even coals.

Carboniferous limestone also occurs in the Isle of Man, and is found throughout almost the whole of the central portion of Ireland.

Some partial replacement of calcium by magnesium (" dolomitisation ") is observed locally in these deposits, giving rise, therefore, to a magnesian lime on burning.

It should be noted that the " Aberdo " limestone of North Wales is of this period, and is quite distinct from the Liassic limestone of the Aberthaw district referred to above.

Other Types

The limestone used locally as a building stone in South Devon (near Plymouth, &c.) belongs to the Devonian; and the next lower formation, the Silurian, yields a grey limestone which is found at points over a broad irregular patch in Shropshire, along the north-east boundary of Wales, and in the southern Lakes district, and which is burnt locally for lime. Only limited use is made of these deposits, however, and as far as existing records show, there is no economically available limestone for lime-burning over large areas of the tracts named.

References

FOR analyses and general information *re* available limestones in Great Britain, and list of principal limestone quarries, *vide* RIVINGTON, *Notes on Building Construction*, Part III (Materials), loc. cit.; GEOLOGICAL SURVEY, Memoir. *The Cretaceous Rocks of Great Britain,* Jukes-Brown, 1903; and HOWE, *The Geology of Building Stones*, London, 1910. A detailed list of varieties of Limestone is given in KNIBBS, *Lime and Magnesia*, loc. cit.

General References

—— *Amer. Soc. Test. Mat. Proc.* Specifications, 1924.

—— *Chem. Met. Eng.*, 1922 (Aug. 23).

ECKEL, E. C., *Cements, Limes and Plasters.* Introduction, loc. cit.

EISEMANN, F., *Baukalk, Kalkverlag, G.m.b.h.* Berlin, 1925.

GEOLOGICAL SURVEY, *Geological Map of British Isles.* 2nd edition, 1912.

Ibid., Memoir. *Cretaceous Rocks of Great Britain,* loc. cit.

Ibid., *Special Report on the Mineral Resources of Great Britain,* Vol. VI, 2nd edition, 1920, Chapter XIV, 190; Vol. XVI, Chapter IX, 79. (On Dolomites.)

H.M. INSPECTOR OF MINES, List of Quarries under Quarries Act.

HOWE, *Geology of Building Stones,* loc. cit.

KNIBBS, N. V. S., *Lime and Magnesia,* Chapter 1, loc. cit.

MITCHELL, *Building Construction, Advanced Course,* loc. cit., Chapter 1.

NAT. LIME ASSOC. (U.S.), *Pamphlets, Construction,* Chapter I, 1923.

OCCHIPINTI, *Giorn. chim. ind. applicata,* 1923, **6**, 17. (Magnesian Limes.)

RIVINGTON, *Notes on Building Construction,* Part III. Materials. Chapters I and III, loc. cit.

U.S. BUREAU OF STANDARDS, Circular No. 30, July, 1920. *Lime: its Properties and Uses.* Sect. 3.

Ibid., Circular No. 106, Nov., 1920. *Lime: Definitions and Specifications.*

Ibid., Circular No. 201, Feb., 1925. *Master Specification for Quicklime for structural purposes.*

Ibid., Technologic Paper No. 16, Feb., 1913. *The Manufacture of Lime,* loc. cit.

VICAT, L. J., *Treatise on Calcareous Mortars.* Trans. Smith, 1837.

SECTION IV.—THE USES OF LIME IN BUILDING

The Slaking of Lime

BURNT lime or quicklime is seldom used as such in building operations, but is practically always slaked before use, and then mixed with sand, hair, &c. The actual slaking process will be discussed in greater detail in connexion with plasters and mortars ; here it is sufficient to state that "slaking" is caused by the formation of hydroxides of calcium and (to some extent) of magnesium by the action of water on the oxides ; the calcium oxide reacting more quickly and evolving more heat than the magnesium oxide if the conditions of burning and subsequent storage, &c., have been the same. It is carried out by the addition of water to the quicklime, and is generally accompanied by the liberation of a considerable amount of heat, and a consequent disintegration of the lumps of quicklime, as described in connexion with Vicat's classification tests. As will be shown in the following Sections, the method of carrying out this slaking process should vary according to the type of lime and the purpose to which it is to be put. For the present, the previous slaking of the lime or the use of a mechanically-slaked hydrate is assumed. It should be noted, however, that "air-slaked" lime is a quicklime that has taken up carbon dioxide, in addition to water. The particles affected are chemically indistinguishable from finely ground chalk or limestone ; and the lime is rendered useless in building operations.

The practical uses of lime, which are of the most interest for building purposes, are in connexion with limewashes, plasters, mortars and concretes.

Limewashes

Under "limewashes," besides whitewash, distemper, calcimine, &c., there must be included, at the present time, some mention of the cold-water paints having as their essential cementing material a calcium caseinate or other calcium compound of glue-like properties, made up locally and on a small scale with lime, the latter slaked on the spot or procured in the hydrated condition ; as well as of the simple limewashes mixed with water or thin size (glue).

Plasters

Under "lime plasters" it will be necessary to deal not only with the straight lime-sand-hair plasters, but also with the modern practice of gauging a straight lime-putty plaster with proportions of rapid-setting plaster-ingredients such as calcined gypsum (plaster of paris), &c. ; but not with plasters consisting essentially of substitutes for lime putty, such as straight plaster of paris (commonly referred to simply as "plaster"), or the several modern artificial substances, sold under traditional or proprietary names. The latter include such materials as Keene's, Parian, Sirapite, &c., generally based on burnt gypsum with various admixtures, such as alum, borax, &c., which are intended to control the rate of setting and are in some cases of the nature of trade secrets, not necessarily identical with the materials indicated in the usual works of reference.

Lime Concrete

Under "concrete" are included mixtures of sand and larger aggregate with hydraulic limes, and of the same with high calcium limes and pozzolanic materials, rather than the poor concretes obtainable with high calcium limes alone without hydraulic constituents; and excluding Portland or natural cement concretes with lime additions; the controversial point as to the effect of the addition of hydrated lime to Portland cement concrete being outside the scope of this Report.

Other uses of Lime in Building

The use of lime in the manufacture of sand-lime bricks; the addition of lime in small proportions to cement-stucco wherein the essential cementing material is artificial Portland cement; the manufacture of "natural cement" by actually *sintering* an "Eminently Hydraulic" (Class C 3) lime (or mixture of already burnt lime with other ingredients) in a kiln identical in principle and operation with the Portland cement kiln; these topics will not be dealt with in this Report. "Roman," "Medina," "Selenitic," and other proprietary cements are included in this category.

Similarly, gauged plasters in which slaked lime is not the predominating ingredient will receive but brief mention here.

The manifold applications of lime in the chemical and other industries are also excluded.

References

KNIBBS, N. V. S., *Lime and Magnesia*, Part III, Chapter III, loc. cit.

U.S. BUREAU OF STANDARDS, Circular No. 30, 1920. *Lime: its Properties and Uses*. Loc. cit.

SECTION V.—LIMEWASHES

A MILKY suspension of hydrated lime in a saturated solution of the lime in water has been in use by man from time immemorial, for the purpose of the decoration of his dwellings. In addition to an æsthetic effect the mild antiseptic properties, as well as the sanitary influence of the clean, bright surface produced, have been long valued. Whitewashing is thus a primitive, but often a quite effective, sanitary measure. In addition, it has a definite preservative effect on metal and other surfaces. In particular, the production of rust on exposed iron or steel surfaces is discouraged by the presence of lime.

Plain Lime and Lime-Salt Washes

The simplest "whitewash" consists merely in a milk of slaked lime in water; this, however, easily rubs off, and disappears almost entirely if exposed freely to the weather for some months. Another simple type of whitewash is made of whiting—finely divided calcium carbonate or chalk—suspended in thin glue or size-water; but as this does not contain lime it is not dealt with in this Report. A plain limewash can be greatly improved by the addition of common salt, in the proportion of 15 lb. common salt to 50 lb. of hydrated lime or the putty obtained by careful slaking of half-a-bushel of lump quicklime, with 7½ gallons of water. If available, 5 lb. of dry calcium chloride (crude commercial) may be substituted for the common salt, and produces a cheap and strongly adherent wash. Either of these salts increases the solubility of the lime, and the degree of penetration; it holds the moisture, and facilitates the carbonation of the lime after application.

Other Additions: Lime Glue-Formula

In a comprehensive series of practical trials of some 121 different whitewash and aqueous lime paint formulæ by G. J. Finck, in which smooth test boards of cypress wood with a black band painted across them under the whitewash (for testing covering power) were exposed to the full effect of the weather and were examined and reported on at the end of 10 days, 30 days, 6 weeks, 2 months, and 3 months, little improvement was observed from the *addition* of a number of materials which appear in current formulæ, such as plain glue, whiting, or flour paste, &c., to the simple lime-salt formula. Lime-glue formulæ usually require a weight of glue about one-twentieth of the weight of lime used, and the same proportion of lime to water as in the simple limewash formula. It was found in this series of tests that when glue was added lime alone gave better results than whiting alone, that the addition of sodium silicate ("water-glass") gave no improvement in permanence, and that the addition of a small quantity of alum improved the working qualities, whilst formaldehyde improved the resistance. For dry interiors, a ***simple non-rubbing formula*** is recommended : 3 lb. of glue dissolved in 2 gallons of water ; and added slowly to a cream made with 50 lb. of hydrated lime or from ½ bushel of lump quicklime carefully slaked and strained through a fine screen after slaking, with 7 gallons of water; thinning as required.

Cold Water Paints

In the cold water paints, most of which are essentially lime-caseinate paints, an adhesive or cement-forming material is present ; an " extender " in the form of excess hydrated lime, whiting, zinc oxide, &c. ; colouring matter if required ; and usually some preservative to prevent putrefaction of the organic constituent before application, or appearance of mould afterwards if applied to a damp wall.

Lime-Casein Water-Paints

In the series of tests by Finck mentioned above, of some forty different substances of this type tested, the results showed throughout the whole series that casein was the best. The value of the casein depends on its ability to unite with the lime to form a compound, calcium caseinate, which becomes insoluble on exposure to air. Solution of the casein is facilitated by the addition of several basic substances, the best of which proved, in these trials, to be (tribasic) sodium phosphate, which is strongly alkaline in solution. This gave better results than ammonia, washing-soda (soda-ash), sodium silicate (water-glass), caustic soda (lye), or borax ; though the last can be substituted for the sodium phosphate in the formula given here without much harm. The best mixture of all proved to be that of lime, casein, and whiting, with the addition of sodium phosphate and formaldehyde (formalin) ; this gives a cold-water paint which will not rub off or chalk, and is quite weather-resistant. Neither common salt nor sodium fluoride improved this, when added in small proportions.

For a cheaper paint, though still of good weather-resisting and non-rubbing properties, the whiting can be replaced by twice its weight of lime. Where borax is used in the place of sodium phosphate this acts as a preservative, and the formaldehyde is therefore omitted.

The casein is the ordinary commercial quality of powdered casein, prepared by one of several available processes (rennet, souring, acid treatment) from separated milk on a large scale ; it is sometimes a little difficult to wet it thoroughly in order to get it into solution, so that it is recommended to soak it in hot water for a couple of hours as a preliminary treatment. The lime cream is preferably made up some hours in advance of use, to ensure thorough slaking and disintegration. If quicklime is used it should be fat, easily slaking lime ; the resulting paste after slaking should be strained through a fine sieve before use, and both casein solution and lime cream should be quite cold before mixing. For small quantities at least the use of ready hydrated lime is recommended. The mixed paint must be used within twelve hours.

The following formula will give a first quality weatherproof wash: soak 10 lb. of commercial casein in 4 gallons of hot water, until thoroughly softened; dissolve 6 lb. (tri-basic) sodium phosphate in 2 gallons of water; add this solution to the casein with constant stirring until all dissolved. Make a thick, smooth cream of 25 lb. whiting and 50 lb. of hydrated lime (or the putty made from ½ bushel quick lime, carefully slaked and screened). Add the cold casein-phosphate solution to the cold lime paste, stirring constantly. Just before use add *slowly* and with vigorous stirring a solution of 5 pints of ordinary commercial formalin (formaldehyde) in 3 gallons of cold water.

Borax can be substituted for both the (tri-basic) sodium phosphate and formaldehyde if the latter are not available; tinting colours can be added as desired. The formaldehyde must be added slowly, with constant stirring, as otherwise the whole may turn to jelly and be ruined.

A rather cheaper cold-water paint results from replacing the 25 lb. of whiting, in the above formula, by 50 lb. hydrated lime. Another excellent lime-casein paint for use on damp walls uses ammonia to get the casein into solution: 5 lb. of casein are soaked in 2 gallons of hot water, and a solution made with 3 pints of commercial strong ammonia solution diluted with a gallon of water is added, and the whole stirred until the casein dissolves completely. Then this is added to the thick cream made with 50 lb. hydrated lime (or by slaking, as before, ½ bushel quicklime) and 6 gallons of water, and well stirred. Just before use, a solution of 5 pints of formalin in 3 gallons of water is added slowly and with constant stirring. The resulting paint can be thinned as desired, and should be used within a day after mixing.

If hydrated lime is used, it is practicable to mix the materials dry, and to make up small quantities with water to give a working consistence; the formalin is added as before, in the same proportion, and just before use.

Coloured Limewashes

A number of pigments are available, which can be added to either of the whitewash formulæ given here. The pigment must be "lime fast," *i.e.*, not affected by free alkali, which will be present in the form of uncombined lime; if other than those directly recommended be used, preliminary test should be made with a small batch to ensure that this is the case. For greys and dark tints, ivory, bone, Paris, lamp, Frankfort and carbon black have been recommended; for a dense white zinc white and lithopone have been added to the lime-caseinate paints. For red there are vermilion, Indian red, Venetian red, and carmine lake. The ochres and siennas are all lime-fast ; for yellow there are available yellow ochre, sienna, and Naples and cadmium yellow. For brown, brown ochre, sienna, umber, Vandyke and Cassel brown ; for blue, ultramarine and cobalt blue are recommended ; whilst for green, emerald green and permanent green have been used. Some caution should be exercised in connexion with lake colours in general ; it is stated that pigments which have been precipitated on a basis of calcium sulphate are to be avoided in limewashes.

Application

Either of the glue or casein preparations can be used as sizing coats on walls before painting or papering. In applying them, the wall should always be thoroughly cleaned by scraping and brushing, any old whitewash, paper, grease and dirt being carefully removed before putting on the new coat. The surface should not be too dry. Quite a thin coat spread on quickly by a large brush, or air-spray, gives the best results. A gallon of the lime-wash should cover about 200-250 square feet (less on brick) ; with a 4-inch brush some 20 to 40 square yards per hour is stated to be a reasonable rate of application. The coat of the more expensive lime-caseinate paint amounts to a few pence a pound only, depênding on local prices for casein and formalin.

A possible development in the direction of the use of a cheaper vegetable protein to replace the casein, such as that from the castor-oil bean, should be noted here ; mention must also be made of some work carried out recently on waterproofing distemper by the addition of gum resins and lime soaps to a lime-caseinate body. Along neither of these lines, however, has a practical result yet been attained.

References

—— *Architects J.*, 1923, **57**, June 13. Washes for Outside Work.

—— *Chem. Trade J.*, 1923, Aug. 31. Casein Paints.
Ibid., 1925, Oct. 30. Casein Paints.

—— E. P. No. 17025, 1921. Improvements in the Manufacture of Coating Compositions.

FINCK, G. J., *J. Ind. Eng. Chem.*, 1922, **14**, 505. Investigation of White-washes and Aqueous Lime Paints.

NAT. LIME ASSOC. (U.S.), Bulletin, 304—B ; *Whitewash and Cold Water Paints ;* and Pamphlets. Construction. 1922. Chapter X.

Specification. Painter and Decorator, p. 531. Architectural Press, Ltd., London (1926).

U.S. BUREAU OF STANDARDS, *Technologic Paper, No.* 181, Dec., 1920. Coloured Wall Plaster, p. 5.

SECTION VI.—LIME PLASTER

PLASTER is defined as any material used in a plastic state to form a durable finishing coat to the surfaces of walls and ceilings, &c., of buildings.

This definition deals only with its use and location of use without reference to its composition. In this Report, however, the term "plaster" has been further restricted to material used for interior surfaces, and only essentially *lime* plasters will be considered.

Plaster is rarely called upon to sustain any great load: it suffices if it is strong enough to support its own weight, together with that of any enrichments forming ornamental work; that is, the keys formed behind the laths must not shear off owing to stresses set up by a moderate amount of vibration or by any subsequent small movements of the building; and if at an early stage sufficient strength is developed to allow of the application of a second or third layer of plaster over the first coat. Since such a moderate amount of strength is easily attained, the most important quality in a plaster is freedom from *cracking* and "pitting" or "popping."

Cracking of Plaster

Large gaping cracks, generally running from corner to corner of a room, or from a corner of a door or window, are probably settlement cracks, due to movements of the building after the plaster is set, and not necessarily due to any defects in the plaster itself. These are dealt with by simply "pointing up" after the building has found its level. A network of fine cracks, going through the plaster and forming large geometric figures on the surface, may result from insufficient bond between consecutive coats of plaster; these cracks are produced by uneven expansion and contraction of the surface coating and backing. Small, fine, superficial cracks over limited areas are produced by shrinkage whilst drying. Lime contracts considerably whilst drying; this is counteracted in plaster and mortar by the addition of sand, and in a finishing coat by the addition, in many cases, of a quick-setting material such as calcined gypsum (plaster of paris) which sets before the lime has had time to dry and shrink much.

The cause of cracks can therefore be generally determined by their nature and manner of occurrence. They are due either to movement of the building, to the use of a badly proportioned finishing coat, or to insufficient or ill-timed trowelling of the finishing coat. Small shrinkage cracks are effectively concealed by papering, and are then relatively harmless.

Pitting of Plaster

Occasionally it is observed that in a plaster surface, apparently free from any defects, and which has been in position for times ranging from months to years, small conical pits will suddenly appear, and fragments may even be seen to be projected from these popholes with considerable violence right across a room. Extended investigations with test-panels made up with plasters to which various impurities had been added have shown that this pitting or popping is due to expansion of particles of certain impurities in the lime, through slow slaking long after the main portion of the lime has set and hardened. The resulting bursting stress on the envelope of hardened plaster surrounding the particle finally releases itself almost explosively, blowing out a small crater in the surface. The offending particle can sometimes be noted at the bottom of the crater. Particles of overburnt lime, iron pyrites, and the products resulting from burning at a high temperature of limes containing clay, iron oxide and silica, are found to produce this effect, if above the size corresponding to a No. 50 sieve. The harmful impurities may be introduced in the sand used.

The *remedies* are : first, the use of a pure, carefully hydrated lime, preferably mechanically hydrated and stored some time before use ; secondly, the removal by screening of all particles that are retained on a No. 50 sieve (hydrated lime, carefully air-separated, will largely conform to this standard) ; thirdly, the use of clean sand. The traditional and simplest method is to allow a lengthy period of seasoning or tempering, thus ensuring the complete hydration of all particles and materials present before actual use. This, however, is probably seldom practical nowadays.

A distinction may perhaps be drawn between the pitting that shows a *dark* particle at the bottom of the pit, which is probably solely due to overburnt or semi-fluxed lime ; and that in which a *light* particle appears, which is merely due to unskilful slaking.

Tempering of Slaked Lime for Plaster

Lime used in plaster-work is generally slaked with an excess of water to form a putty, and this putty is then stored in a covered pit or trough until used. Practically all the authorities, both ancient and modern, are unanimous in demanding an adequate length of storage time for *tempering* this slaked lime before it is actually used. Pliny expresses clearly the classical view-point when he says : " The older, too, the mortar is, the better it is in quality. In the ancient laws for the regulation of building, no contractor was to use the mortar less than three years old ; hence it is that no cracks have disfigured the plaster coatings of their walls."* Vitruvius required that lime should be tempered a long

* " Intrita quoque, quo vestior eo melior.
In antiquarum aedium legibus invenitur
ne recentiore trima uteretur redemptor.
Ideo nulla tectoria eorum rimae foedavere."
—PLINY, *Natural History*, XXXVI.

English version from Bostock and Riley's translation, Bohn's Classical Library, 1855, p. 374. " Trima " is mis-translated as " three months " in the original.

time before it is wanted for use ; most modern writers call for a period of from two weeks to two months. Mitchell quotes Specification Clauses of the London County Council requiring that lime for plaster should be run to putty at least one month before using, and then screened to remove unslaked particles. There appears to be no practical limit to the period of useful tempering ; there was an old French proverb to the effect that " lime at one hundred years is still a baby." Alberti describes how he found in an old trough or ditch some slaked lime which, from various evidence, had been abandoned there more than 500 years before, and which was in excellent condition for use ; in recent times J. D. Batten has described how, by tempering the slaked lime in a covered pit for from eight to eleven years, a material was obtained of particular value for such work as fresco painting on fresh plaster surfaces.

Whilst it is hardly practicable at the present time to afford always so long a time as the traditional three months or a year for the tempering process, it is evident that for good permanent plaster-work great care should be taken in slaking to avoid unslaked particles of appreciable size, and to remove any overburnt residues ; a reasonable time should also be allowed between slaking and actual use. With mechanically hydrated lime, particularly if stored dry some time, and made into putty the day before use, the danger of the presence of large, unslaked particles is very small.

Test for Pitting

In order to determine whether there is any likelihood of pitting or popping with a particular plaster, S. E. Young has suggested a practical test ; this is to spread out a thin layer ($\frac{1}{16}$-in. thick) of the gauged material on a sheet of glass, and hold it above a strong light. If dark spots are noticed, there is a possibility of pitting developing later on.

Magnesian Limes and Pitting

Magnesian limes slake more slowly than high calcium limes, and often a large proportion of the magnesium oxide will escape hydration altogether, particularly if burnt at a high temperature. They expand rather less on slaking, and shrink less on drying, than calcium limes ; they also set more slowly. Some authorities claim that, as a result of this slow setting, there is actually less risk of pitting occurring with a magnesian lime, as any secondary hydration processes have ample time to complete themselves before the plaster is finally set. As magnesian limes in general are of a pure white colour, and exhibit a high degree of plasticity, the writers referred to express preference for these limes for use in plaster-work.

Plasticity

Plasticity, from the plasterer's standpoint, is a difficult property to define. Such simple descriptions as " the property (of lime plaster) which makes it easy to spread " and " the yielding property of a wet mixture to change of form " convey the general idea, but are lacking in scientific precision or quantitative meaning. A formal definition of plasticity in general is : " A property of solids in virtue of which they hold their shape permanently under the action of small shearing stresses, but they are readily deformed, worked, or moulded, under somewhat larger stresses." " Plasticity is thus a complex property, made up of two independent factors, which we must evaluate separately." (Bingham.)

Emley, of the United States Bureau of Standards, in connexion with the specification of their official " plasticimeter," has pointed out that in plasters there is still another factor which affects ease of working, and which must therefore be considered in a definition of practical plasticity ; this is the retentive power of the plaster for water, and the consequent prevention of drying out by the " suction " of a partially dry plaster, masonry, or similar surface on which the plaster is being applied, and which tends to absorb by capillarity into its own pores the water which is so essential a constituent in the plaster mixture. A plaster which is rather " short " but which is very tenacious of its essential water-content will spread farther on a porous thirsty surface than a material initially softer and more " plastic " (according to the more general definition), but which rapidly loses its water and becomes thereby dry and friable. Well-matured slaked lime has this property of water-retention developed to a high degree, and can accordingly be used where other plastering materials fail.

Measurement of Plasticity

Whilst the plasticity of plasters is thus a property which is hard to measure, quantitative measurements are essential for the accurate comparison of different materials, for testing the effect of particular treatment of the lime, for the determination of sand-carrying power, &c. ; so that investigators have considered it worth while to expend a great deal of time and effort to develop a testing machine which will, to some degree, give a numerical record of this property, independent of personal judgment.

The Emley machine is based on the principle that that material is the more plastic which has the great ability to retain its water against the " suction " of the surface to which it is applied ; and that that material is the more plastic which requires the less work to spread it. When the water has been reduced to a certain proportion, the plaster is no longer workable ; and it is evident that a plaster which can be spread with less effort will be preferred in practice, and will allow of better-finished work and quicker application. The difficulty is to apportion correctly the relative weight to be assigned to each of these two factors, as measured in a machine ; and to distinguish with certainty between a sticky, rubbery material which holds water tenaciously, a truly plastic smooth-working putty, and a wet sandy material which spreads quite easily at first, but soon dries out and crumbles under the trowel.

A measure of the " work " factor in plasticity can be obtained by determining the viscosity of the paste ; a plunger of known dimensions is immersed in it, and then pulled out by the falling of a known weight. The viscosity is then proportional to the time required to pull it out. The method is not available for thin pastes. Another method is that of rotating a plunger or paddle immersed in the paste, and measuring the force required to turn it. However, the viscosity alone depends so much on the water content that this method gives but limited information as to the real plasticity.

Resistance to " suction " alone can be measured by spreading the paste on absorbent plaster blocks, and observing the rate of drying out. A very convenient and simple test, based on this but including the other factor as well, is the *Carson Blotter Test.* This requires no more apparatus than a piece of blotting paper, a tin cup, and a table knife. The lime is mixed with water in the cup to a good consistence (a slight variation is unimportant) ; preferably left overnight to temper ; and then spread like butter on the blotting paper with the knife (or spatula). With a little experience, and with a standard lime of known good plastering qualities for comparison alongside that under test, by noting how the material works and how far it will spread on the absorbent surface, quite good and reliable estimates can be made, and small differences in plasticity observed with certainty.

Emley Plasticimeter

In the final Emley plasticimeter, developed after a number of years' work in investigating limes, and the trial and abandonment of a number of earlier designs, the " suction " factor is provided by a porous plaster block on a turntable, on top of which the sample of lime paste is to be placed. The paste is moulded into a cylindrical disc 3 in. in diameter by 1½ in. high. The consistence of the mix may vary somewhat ; it should be a good workable paste, and can be defined more accurately by a slump test. A plaster of normal consistence is taken to be such that a cylinder moulded 2 in. in diameter and 2½ in. high slumps ½-in. when the mould is removed.

The " work " factor is measured by an indicator, which shows the torque developed when this disc of paste is very slowly screwed up against a very shallow conical disc held stationary above it. This disc is intended to correspond to the plasterer's trowel, and the slow rotational movement and upward creep to the action of spreading plaster on the absorbent surface below. Actually the turntable rotates once in 6½ minutes, and rises one-thirteenth of an inch in the same time. Curves are plotted of torque developed and of time; the area under the curve gives a measure of the *work* done in spreading the paste; the point at which the paste has dried out so that it ruptures, *i.e.*, is no longer capable of plastic flow, is shown by a break in the curve; and the time elapsed before this rupture occurs gives a definite measure of the *resistance to "suction."*

Classification for Plasticity

By means of this machine limes can be accurately placed in the "5, 10, 15, 20, 25 and 30-minute" classes, according as to

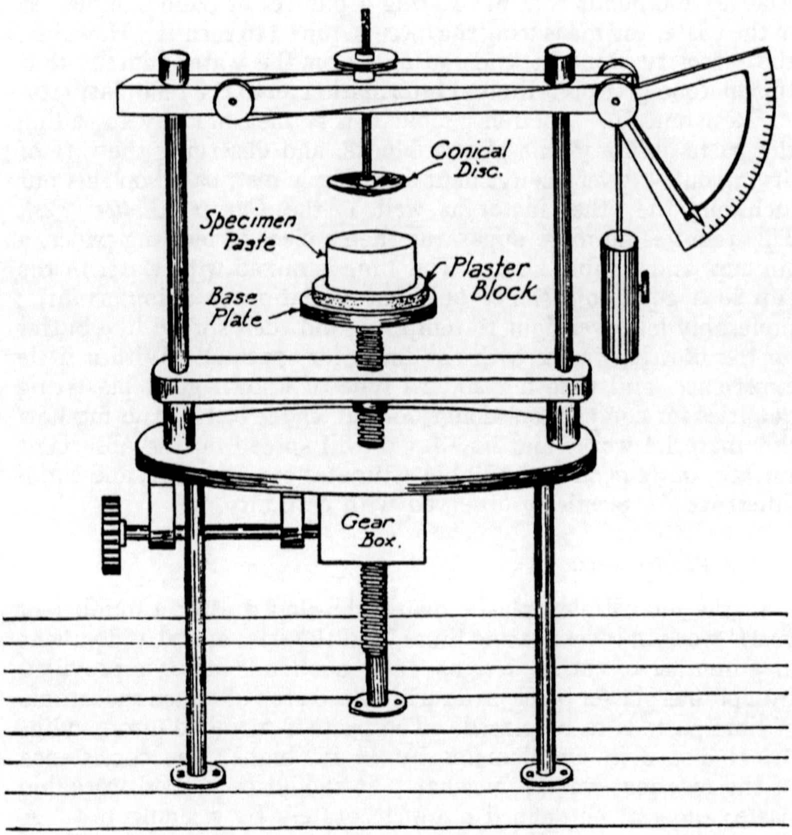

FIG. 3.—Sketch of Plasticimeter.

whether the rupture occurs after the first 5, 10, &c., minutes in the machine; and each class further subdivided according to the work expended in spreading them during this time. The longer the time, and the less the work, the higher is the practical plasticity of the lime. Hence a numerical meaning is given to the rather ill-defined quality of "plasticity" and a permanent record, independent of any personal judgment or prejudice, is obtained for each specimen tested. Accordingly the instrument is used practically to determine the plastic qualities of limes in official specifications of limes for use in plaster work, a definite plasticimeter figure being required.

It is obvious that the Emley plasticimeter is open to many criticisms ; but it represents the most practical solution of the problem of *measuring* this quality to date ; the only alternative is that of a practical trial on a small scale in actual plastering over a typical backing of laths, &c., by an expert craftsman, and subject therefore to a large personal equation, and to a complete lack of any quantitative records.

Sand-carrying Capacity

On account of the marked contraction accompanying the process of drying and setting of neat lime plaster, it is essential to add a proper proportion of sand to the plaster for the thicker under coat or coats. Also for economical reasons the greater the proportion of sand the plaster can carry without deteriorating the better ; the sand-carrying capacity of the lime is therefore of considerable importance.

It is evident that shrinkage will be almost entirely prevented when the sand grains (which do not alter in volume) are in actual contact with one another. The spaces in between the grains will then be filled wholly or partially by fine particles of lime hydrate and (at first) limewater, unless indeed the hydrated lime contains a large proportion of aggregated particles of a size comparable with that of the sand, so that they are too large to fit within the pores. With a poor or lean lime there will be present a certain proportion of finely divided clay, &c., to supplement the finer sand particles. There should just be enough, but not more than enough, of the cementing material, the lime, to fill the voids. Excess of lime would produce shrinkage of the whole mass ; deficiency of cementing material would result in some loss of strength. Too fine a sand will not allow of a proper distribution of the lime, as the pores will frequently be too small in comparison with the size of the hydrated particles, so that the strength will suffer. Accordingly, the sand-carrying capacity depends not only on the plasticity of the lime (which is closely related to the size of the hydrated particles) and upon the amount of impurities present in the lime (also affecting the plasticity), but also on the average size and distribution of sizes of particles in the sand used. It will thus be affected by the method of slaking used, and upon the amount of maturing of the putty, since these factors influence particle-size and resulting plasticity. In practice, sand would be added until the plasterer began to find an appreciable difficulty in working the material, on account of decreasing plasticity. In general, this will imply rather more sand than theoretically required to give optimum conditions, but the resulting strength will not necessarily fall short of the moderate requirements for ordinary plaster-work. Specifications usually include the demand for a definite proportion of sand to lime for each coat ; for the best and most lasting work the type of lime, mode of slaking, and grading of sand should also be mentioned.

The Setting of Plaster

At least two stages, possibly three, must be recognized in the setting of lime plaster. Broadly speaking, there are the first stage of drying out, common to any type of plaster irrespective of its chemical nature, during which excess water is evaporated from the thin layer on the wall; and, secondly, the subsequent hardening, by chemical action between the calcium hydrate and the carbon dioxide of the air, resulting in the formation of calcium carbonate and the liberation of a further quantity of water, which evaporates. To the formation of a coherent mass of crystals of calcium carbonate—practically limestone—is, then, attributed generally the strength of old plaster.

However, as will be shown at greater length in connexion with lime mortars, this process of carbonation is excessively slow and generally incomplete, only a shallow surface layer being affected in any reasonable length of time. Minute traces of quicklime were found in the pure lime plaster on the walls of the palace of Knossos, in Crete, which had been exposed freely to the atmosphere for 3500 years; and free lime is found in the plaster used in the Pyramids. German investigators have found that the useful penetration of atmospheric carbon dioxide is but $\frac{1}{25}$ in. in interior plaster-work.

Another stage can be recognized in the setting of a plastic lime plaster, which is included broadly in the first or drying-out stage mentioned above.

This is the slow partial solidification of lime putty after "knocking-up."

"Knocking-up" of Lime Putty

A property peculiar to slaked lime putty, and particularly noticeable in putty which has been left a long time in a covered pit or trough to mature, is its power of forming a homogeneous mass of slight but definite rigidity when a certain amount of the contained water is allowed to evaporate or drain away. Very slight stirring or working—"knocking-up"—often without the addition of water, suffices to break up this structure so that the mass as a whole resumes its high plasticity. This phenomenon has been studied recently by J. D. Batten and the late Sir George Beilby, F.R.S., and various suggestions have been put forward to explain the observed facts. Investigators have attempted to follow closely the processes involved in the action of the slaking water on the surfaces of the lumps of quicklime (both external surfaces and those of the pores in the material); and those involved during the subsequent tempering of the lime putty produced.

According to the work of Kohlschütter and others, fresh lime putty can be looked upon as made up of a large number of minute particles, each separated from its fellows by its own little surface envelope of milk of lime in the form of a gelatinous emulsion (Bur. Stand. Cir. 151, p. 28). These minute particles, of colloidal size, result directly from the interaction of the water and the quicklime at the surface of the latter, whether on the outer surface or in the pores of the burnt stone. These particles may give rise, then, by coagulation to secondary aggregations, so that the putty may consist either of separate primary particles, or of aggregations of varying size formed by the primary particles, the whole suspended in a solution of calcium hydrate. The plasticity or working qualities of the putty (and hence of the plaster made with it) will evidently depend on the relative proportions of the primary and secondary states of division which depend in turn on the mode, temperature, and rate of slaking. For minute colloidal particles will readily flow when the putty is worked, whilst large irregularly shaped aggregated particles or clumps would give rise to an increased resistance to flow, and produce a " short " material. The effect of subsequent maturing (" ripening," " curing," " soaking," " ageing," &c.) of the putty over long periods can be attributed to a further process of colloidal dispersion of the larger, aggregated particles, producing thus a finer, more mobile, and uniform material of high plasticity.

The subsequent partial hardening of the finely-dispersed mass was attributed by Sir George Beilby to a molecular association set up between the lime hydrate and previously free water molecules, but of such a temporary and feeble character that even mechanical disturbance is able to break it down. According to this theory, then, the plasticity of lime putty results from the transient flow of the molecules under mechanical disturbance and the restoration of rigidity when the disturbance ceases ; this process being repeated as often as the putty is disturbed. Continental investigators have attempted to assign actual chemical formulæ to such complex hydration products, and to explain these in terms of " secondary valencies." An alternative, purely colloidal, theory likens the structure of the lime putty to that of a jelly, which simply dries out to a hard mass when the plaster sets, after the manner of gelatin.

The regular beating or chopping up of the mixed plaster so frequently referred to in the history of plastering, by Vitruvius and subsequent writers, and met with in the traditions of fine plaster-work, will evidently have the effect of promoting the deflocculating or dispersive process in the slaked lime, with a resulting gain in plasticity. A similar process is that suggested by F. C. Welch, of grinding a hydrated lime of but moderate plasticity (obtained by slaking lime at a high temperature and containing flocculated particles) with a small amount of a desiccating agent, such as quicklime ; it is found that a highly plastic lime is thus obtained.

Drying of Plaster

An important factor, affecting both the immediate inhabitability of a building and the durability of the plaster itself, is the proper rate of drying of plaster. Too rapid drying may be a source of failure in plaster-work which may be correct otherwise both as to composition of the material and method of application. This may be due either to lack of precautions in hot weather, or to improperly applied artificial heat during cold weather.

The preliminary hardening of the plaster takes place, quite rapidly, by the loss of water through the " suction " of the surface on which it has been applied and through evaporation ; during this period occurs the spontaneous setting of the putty mass discussed in the last paragraph. Following this, the loss of water due to evaporation becomes more gradual ; and there is a subsequent partial and superficial carbonation (in any reasonable length of time) by atmospheric carbon dioxide, as indicated just above. The water produced during the last chemical process (*i.e.*, carbonation) has to be evaporated, but represents a relatively insignificant amount during the period under consideration.

The water added to a plaster during mixing is merely for the purpose of rendering it plastic and so in a fit state to be applied. It has no chemical effect, and once the plaster is in position this water must be removed ; but not so quickly or violently as to cause sudden volume changes, *i.e.*, sudden shrinkage or powdering. It is otherwise with gypsum and Portland cements ; with the first, a considerable amount of water (almost half of the normal proportion of gauging water) is required for completing the chemical process of setting ; and with Portland cement also a considerable amount of water is required for this purpose over a lengthy period, so that it should actually be maintained in a damp condition for some time after application.

In lime plaster, it has been explained how the sand, as well as being an economic diluent or extender, acts so as to prevent cracks developing through the comparatively large volume changes involved in the drying out and setting of the lime putty. There is necessarily a gradual process of self-adjustment to this shrinkage ; extremely rapid drying may make difficult this process of settling down, resulting in the development of cracks.

Another, though less important, point is in connexion with the carbonation of the lime. Various investigators have observed that little absorption of atmospheric carbon dioxide takes place until mortar is fairly dry in the first instance, but not *completely* dry ; Burchartz has found that this process of carbonation is actually greatly accelerated by subsequent periodic wetting of the mortar. When plaster is completely dried out very rapidly the conditions are unfavourable for a subsequent carbonation, and but little mechanical strength will be developed.

Accordingly, the drying should be moderately slow and general, and any intense *local* drying effects should be avoided. Experience shows that the *circulation* of the air is more important than either the temperature of the air, or its degree of humidity ; the limiting factor in any case is the slow process of diffusion of water-vapour through the pores of the plaster mass.

Strong, direct draughts on any portion of the surface should be avoided, for the reasons given above ; the provision of fabric coverings to window and door-openings is therefore a desirable course, allowing proper ventilation without strong air currents. When extremely cold or moist conditions prevail, and artificial heat is used, this should be looked upon rather as a means of inducing ventilation than as a source of direct heat, and care should be taken to keep the steam pipes, salamanders, &c., well away from any fresh plaster surface. Where steam pipes and radiators are already installed in a building, the incautious use of these to accelerate drying is likely to cause cracking and failure in the plaster closely adjacent, and particularly in that just *above* the fitting. For in such cases both extreme local heating and powerful drying currents of air will be brought into play, and portions of the plaster will inevitably be dried out far too quickly ; this may even be carried so far that steaming is produced, and complete disintegration of the plaster may result.

With artificial heat, then, precautions must be taken to avoid too intense or local heating ; to encourage circulation of the air and adequate ventilation for the escape of the water evaporated ; and (in the case of portable heaters) for the removal of the products of combustion. There is but little point in the suggestion that the presence of a salamander, by supplying a large amount of carbon dioxide, favours in this manner the carbonation of the lime. For whilst the plaster is still wet enough to call for artificial heat to hasten the drying, it is not in a favourable condition for extensive carbonation.

With gypsum plasters (including the various proprietary plasters consisting essentially of calcined gypsum) even greater care must be taken not to remove the water too hastily, since, as a portion of this water is necessary for the chemical process of setting, the material may become friable and disintegrate far worse than a lime plaster would if dried prematurely under similar circumstances. Equal care should be taken with plaster material prepared by gauging lime putty with a considerable proportion of some gypsum plaster, for the finishing coat ; particularly when this is applied over a well-dried undercoat, since the bond will often be found to have failed completely after a short period if the drying out has been hastened unduly by the use of artificial heat. Reference has already been made to the necessity of actually maintaining a Portland cement plaster in a damp condition for some time after application.

Slaking of Lime for Plaster

As indicated in the paragraph on plasticity, the value of the lime as a plastic material is closely related to the ultimate structure and size of the particles of calcium hydroxide which make up the putty mass. Any factor which affects this fine structure will affect the behaviour of the lime when made into plaster. On account of the high plasticity and sand-carrying powers of fat high-calcium (or magnesian) Class A limes, these are largely used for plastering ; so that we have here to deal only with the slaking of such limes.

There is a danger of burning a fat lime in slaking, as a result of an insufficient supply of water or inadequate stirring. Lime slaked at too high a temperature has a tendency to coagulate the minute (colloidal) particles which are the first product of the interaction of the water with the surface of the quicklime, thus forming aggregates and diminishing the plasticity. The use of too large a quantity of water—" drowning "—in the slaking process is usually considered to have an unfavourable effect, besides giving more trouble by requiring the excess of water to be run off or evaporated subsequently before use.

The method recommended for slaking the hot, high-calcium plastering limes is accordingly that of immersion, in sufficient water to yield ultimately a putty, and with close attention and constant stirring; more water is added later if, necessary, but quite slowly ; and after the violent action and steaming have subsided the whole should be thoroughly worked in order to ensure that all lumps have been broken down.

The milk of lime resulting should then be run through a sieve of not more than $\frac{1}{4}$-in. mesh (preferably smaller) into a large trough or pit to settle, and covered up lightly, the excess of water being allowed to evaporate or drain away. Then the putty formed is left to temper for as long a period as possible, preferably for not less than a month.

With the slower of the magnesium limes, the water may be added to the lime, and less is used at first ; whilst in order to conserve the heat evolved the water should be added slowly at first, so to encourage a fair rise of temperature in the process, which favours the chemical action. Thorough larrying, screening, and a period of tempering are also called for, as with the high-calcium limes

With mechanically-slaked hydrated lime, simple mixture with water, followed by a soaking period of a day, suffices.

Strength and Adhesion of Plaster

As already indicated, the strength of ordinary lime plaster is sufficient for ordinary purposes when an amount of sand is used with it that does not reduce the plasticity below a workable degree ; so that the strength is not a critical matter when the plaster is properly mixed and skilfully applied. The common practice of adding hair (or other fibrous matter) to the lower coat or coats is chiefly a matter of convenience to prevent the plaster from breaking away whilst still wet from the key behind the laths, and to facilitate application generally ; the normal wastage from droppings is also diminished.

Similarly, the adhesive power of the plaster is adequate when the standard specifications as to scoring or scratching the under-coat to provide a key for the next coat, and proper trowelling of a finishing coat, as well as the damping down of too dry surfaces before application, &c., are followed, so will not be discussed further here.

Efflorescence, &c.

If soluble salts are present in the plaster, such as calcium chloride or magnesium sulphate, they will tend to be concentrated at the surface as the water dries out of the plaster. If they are hygroscopic, they may then have the effect of maintaining a damp patch on the plaster, producing blotches ; or in a dry atmosphere the salts may crystallize out at the surface, giving a feathery white incrustation. Even if washed off, the effect may recur until at length all the salts are removed. This efflorescence may easily happen if sea or estuary sand is used in the plaster ; or if very impure, mineral-containing water is used in gauging the plaster. This trouble will not occur if clean pit or river-sand, and pure water are used.

Other faults sometimes met with are hollowness, resulting from imperfect bond between coats ; and peeling of paint and paper, generally traceable to water-leaks rather than to any particular fault in the plaster itself.

Gauging of Plaster

It has become customary, in order to economise time, to add a varying proportion of a quick-setting component, such as calcined gypsum (plaster of paris), or Portland cement, to lime plaster, particularly for the finishing coat. This addition has also the effect of forestalling the normal shrinkage of lime plaster whilst drying out, by making the whole set long before the latter process can proceed any distance, so that the laborious hand-working necessary to give a perfect finish with an unsanded or lightly sanded lime-putty, reference to which is frequently met with in the history of the craft, is largely obviated. For only by skilful trowelling at exactly the right moment in the drying process can the effects of the large contraction of a *neat* or rich lime plaster be removed. The exact proportions of these ingredients which are added are so much a matter of local conditions, and therefore of individual judgment of the experienced craftsman, that it would serve no useful purpose to give any figures here. Sufficient is added to cause the product to set within the required time, the amount being determined by previous experience or actual trial.

Mention should be made also of the expedient adopted where plaster would be subjected to unusually severe wear, as at corners of corridors, &c. ; at these points Keene's or other hard and quick-setting plasters are used, joined up with ordinary straight lime plaster over the main part of the walls.

The Preparation of Surfaces for Plaster

The surface to which lime plaster is applied must provide sufficient bond to support the wet, plastic material when fresh, and to give an adequate key when dry, and must permit of a fairly uniform thickness of coat. The water-absorption power or " suction " of the surface must also be of a reasonable order, so that the plaster shall retain its plasticity whilst being applied. For if the surface is too smooth and non-porous, the plaster will not easily remain in position, and will readily break away when set. Also since the function of the plaster is generally to provide a smooth external finish, if the underlying surface is very irregular, there will be thin places and sudden changes of thickness in the plaster coat, both provocative of cracks and weakness in general. It will be impossible to apply a smooth adherent coat to a surface which drinks up immediately the water upon which the plaster depends for its plasticity, or, alternatively, to a surface from which the plaster slides because it is so wet.

Types of Plaster Work

In ordinary lath-and-plaster work the bond is provided by the proper spacing of the laths, generally about ⅜-in. apart, so as to allow of the formation of an adequate key when the (haired) plaster is applied with some pressure, the hair or fibre usually incorporated in the first coat preventing the ridge of material protruded at the back from dropping whilst still soft. Excessive " suction " of dry laths is combatted by a moderate damping of them before starting the work, excessive moisture (as shown by the visible presence of drops of water) being avoided. In order to maintain the key, it is specified that where studding, &c., is more than 2 to 4 in. wide, " brandering " or " counter-lathing " should be used. This consists in nailing a narrow strip of wood, such as a lath, to the studding, upon which the ordinary laths are nailed. For a similar reason the heading or vertical joints in the laths are recommended to be broken at short intervals. In order to provide a ground work of sufficient strength to admit of subsequent coats being applied without damaging the key of the undercoat, a lath of definite strength must be used. Specifications often call for " lath-and-half," *i.e.*, about 1½ in. wide by about ¼ in. thick. In some districts laths not more than 1¼ or 1⅛ in. wide are commonly used for the best work. Studding or joists should be fixed not more than 12 or 13 in. apart, and each lath should be nailed thereto at every crossing place. The ends of the laths must not be lapped, but must be nailed " single ends," *i.e.*, with butt joints. In a specification for first-class work the proportion and quality of hair or fibre added in the first coat may also be laid down ; and the size of nail used may be specified.

It is often stated that a riven fir lath gives better work than one that is sawn. From the point of view of mechanical strength in the lath alone, the continuity of fibre in the riven lath should make the latter more desirable than a sawn lath ; however, the mechanical strength of a lath is rarely tested very severely, whilst from the point of view of uniformity of bond, and of thickness of the plaster layer (which has to conceal any irregularities in the laths as well as unevenness in the wall itself), the sawn lath may have a considerable advantage. Any sudden local variation of thickness in the plaster layer, to compensate for a crooked or twisted riven lath, will introduce dangers of cracking in the finished plaster.

With metal lath, the bond is provided by the perforated or other particular structure of the lathing; and, being non-porous, the question of " suction " does not arise. Mechanical strength is provided for by specifying a certain weight per square yard, or gauge and mesh in the case of wire lath. The spacing of the studding, &c., and the provision of narrow strips under the lathing where the studding or joists are of considerable width, should be specified as with wooden laths. Similarly, adequate and sufficiently frequent fastening to the supports at each crossing point, and tying of edges of sheets between supports, should receive attention.

A brick or masonry surface is usually rough and porous enough to give an adequate bond without further attention if the joints have been thoroughly raked out. With concrete, it may be necessary to roughen the surface artificially if smooth forms have been used ; but in all cases the surfaces must be damp, and grease spots and laitance must be removed. Vitrified brick and glazed tile give little or no bond, and require either a specially prepared surface or else separate lathing.

In some cases, metal lathing and nails are galvanized or enamelled to prevent corrosion ; this is less likely to occur with a straight lime plaster than with a gypsum plaster.

Whilst the detailed technique of the plasterer's craft lies outside the scope of this Report, a brief description of the different types of utilitarian plaster-work in common use is included. In general, good plaster work on walls or ceilings consists of either two or three coats. Only in the roughest work is one-coat work on laths used : " lath and lay" or " lath and plaster one coat." For rendering on brick or masonry, in a single coat, Portland cement would be preferred to lime plaster.

Those cases where Portland or Keene's, Parian, Martin's, Sirapite, &c., cements are used alone, and where their greater hardness and speed of setting allow of a different technique, are necessarily excluded from this Report. When lime plaster is used there are the alternatives of two-coat work (" lath, lay and set " or " lath, plaster and set "), and three-coat work (" lath, plaster, float and set " or " lath, lay, float and set ") respectively. Similarly, on brick or masonry surfaces, there are the alternatives of rough rendering and setting, or rough rendering, floating and setting respectively.

Each coat may be allowed to become quite hard and dry before the application of the next, as in the traditional fine many-coat plaster work of the Classical and Renaissance periods, and with the best modern work in lime plaster ; or a second coat may be applied whilst the first coat is still moist, but firm enough to resist the force required (" doubled-up work ").

The first, " pricking-up " or rough-rendering coat is composed of " coarse stuff," *i.e.*, sand, lime and, when on lath, hair, with a volume-proportion of about 3 to 1 for the first two ingredients, or 2 to 2½ to 1 for ceilings. The lime need not be one which develops the maximum plasticity for this or for the second coat of three-coat work. The plaster is applied with a fair amount of pressure, so as to make a good bond with brick or other masonry surfaces, or, in the case of lath-and-plaster, so as to squeeze a sufficient proportion through the $\frac{3}{8}$-in. gap between the laths to form a good key. The coat is made about $\frac{1}{4}$-in. thick at first and, except in doubled-up work, is allowed to dry before applying the second or "floating coat," being well scratched (with pointed laths or a coarse broom) before actually setting, to provide a key for the next coat. In doubled-up work, a second coat of coarse stuff is applied immediately, and brought up to about $\frac{5}{8}$-in. thickness (measured from back of laths, as determined by wooden guide strips, known as "grounds," or by plaster "screeds" previously set up) and made true and plane by use of the "rule," "derby," and "float."

In three-coat work the "floating coat" is applied over the scratched and now dry first coat, and brought up to about $\frac{7}{8}$-in. "grounds" as described for doubled-up work. The coarse stuff for this coat may be identical with that for the pricking-up coat; in some cases rather more sand is used (*e.g.*, 3 : 1 in the place of $2\frac{1}{2}$: 1 used for the first coat), and the hair may be omitted.

The finishing or "setting" coat, of "fine stuff," is applied in a thin layer over the floating coat when the latter is dry, and can be omitted altogether in less exacting work. Its function is principally ornamental. An extremely high plasticity, and a regulated time of setting, are essential in the material used for this final coat; it has become customary therefore to combine in the "fine stuff" the high plasticity of a fat lime (Class A) putty with the quick-setting power of calcined gypsum (plaster of paris), by gauging the putty with a variable amount of the latter; and a number of quick-setting substitutes for lime plaster have also come into use. Sand is frequently omitted in this setting coat, and it is to be noted that the ancients used marble dust here. Owing to the absence of sand, and in order to avoid shrinkage-cracks, the coat has to be consolidated by trowelling the surface whilst still just short of setting; as this is quite a critical matter, considerable experience and skill are required of the craftsman to judge exactly the right moment and to use the wet brush and trowel in precisely the right manner to achieve the desired result. Different types of finish can be imparted at this stage by varying the procedure.

The technique of the running of mouldings, and of decorative plaster-work in general, cannot be discussed here, beyond noticing the modern tendency to abandon lime for calcined gypsum or other quick-setting plasters in such work.

References

ALBERTI, *Architecture*, Books II and VI (Historical).

—— *Amer. Soc. Testing Materials Proc.*, 1924, **24**, 303. Appendix. (Preparation of Lime Putty.)

—— *Amer. Soc. Testing Materials Proc.*, 1924, **24**, 850. (Definition of Plaster.)

BATTEN, J. D., *Soc. of Mural Decorators and Painters in Tempera.* Papers Vol. II, p. 82 (loc. cit.).

BEILBY, SIR G. Ditto, p. 125.

BINGHAM, *Fluidity and Plasticity.* London and New York (1922).

ECKEL, E. C., *Cements, Limes and Plasters* (loc. cit.).

EMLEY, *U.S. Bureau of Standards, Tech. Paper No.* 169 (1920). Measurement of Plasticity of Mortars and Plasters.

EMLEY and BERGER, *Amer. Ceramic Soc. J.*, 1923, **6**, 1007 } Pitting of Plaster.
EMLEY and YOUNG, *Amer. Ceramic Soc. J.*, 1920, **3**, 877 }

ENDELL, *Tonind. Ztg.*, 1915, **39**, 13, 73, 85, 107.

HOLMES and FINK, *U.S. Patent No.* 1554183 (1925). (Production of quick-setting lime products by the addition of an aluminium fluoride).

—— *U.S. Patent No.* 1554184 (1925) (Ditto of Zeolites).

JOHNSON, *Materials of Construction*, 5th edition. London and New York (1919) : Withey and Aston. (Magnesian Limes, p. 364.)

JUSTIN–MUELLER, *Rev. gén. colloides*, 1925, **3**, 73. (Colloidal Theory of Slaking Process.)

KEIM, A. W. (trans. Salter), *The Prevention of Dampness in Buildings.* London, 1902. (The Artificial Drying of New Houses, etc. Chapter IV.)

KEMP, W., *The Practical Plasterer.* London (1926).

KOHLSCHÜTTER, *Zeits. ang. Chem.*, 1925, **38**, 305.

KOHLSCHÜTTER and FEITKNECHT, *Kolloid Zeit.*, 1925, **37**, 247. (Colloidal Dispersion in Slaking Process.)

KOSMANN, *Tonind. Zeit.*, 1920, **44**, 555, 565, 632. (Theories of Slaking and Tempering Processes.)

MILLAR, W., *Plastering* (loc. cit.).

MITCHELL, *Building Construction, advanced course*, Chapter IV (loc. cit.).

NAT. LIME ASSOC. (U.S.) *Pamphlets, Construction* (1923) (loc. cit.) (Chapter II, Lime Paste ; Chapter V, Lime Plaster ; Chapter XVII, A. Bull. No. 305A, Standard Specification for Lime Plaster.)

RIVINGTON, *Notes on Building Construction*, Part III, Chapter III (loc. cit.).

—— Specification (1926), *Plasterer*, p. 357.

U.S. BUREAU OF STANDARDS, *Circular No.* 30 (1920), Lime, its Properties and Uses (Sect. III, V), loc. cit. *Circular No.* 151 (1924), Wall Plaster. (Report of Bureau of Standards Plastering Conference). Tech. Paper No. 16 (1913), loc. cit. The Manufacture of Lime (p. 15).

VITRUVIUS, *Architecture*, Book VII, Chapters II, III (loc. cit.).

WHITMAN, W. G., and DAVIS, G. H. B., *J. Ind. Eng. Chem.*, 1926, **18**, 118. (The Hydration of Lime.)

YOUNG, *Amer. Ceramic Soc. Trans.*, 1913, **15**, 659. (Pitting of Plaster.)

—— *Zement*, 1926, **15**, 71. (Technische Vorschriften fuer Bauleistungen, German Specifications for Plaster-Work.)

SECTION VII.—POZZOLANAS

In the Historical Section, I, reference was made to the ancient habit of the Romans of adding a material, pozzolanic sand, to their lime mortars, in lieu of or in addition to the ordinary sand ; and to the comparatively recent revival of the use of such additions, not only to lime mortars, but also to Portland cement mortars, in Germany, Holland, France and Italy.

Whilst but little attention has been paid to this subject in Great Britain since the artificial hydraulic cements became common, great progress has been made in other countries in connexion with such admixtures, and a considerable literature has accumulated, largely in German and French technical journals, on this topic Much experimental work has been carried out recently in these countries, more particularly in connexion with locally available pozzolanas ; so that precise information is beginning to be available for their engineers and builders, and formal specifications have been published for particular materials, such as trass ; whilst important works have been successfully completed on the largest scale in lime-pozzolana and cement-pozzolana concretes, particularly in Western Europe. In addition to improved resistance to sea-water, a very sensible economy has often been effected by the use of a pozzolana.

Definition

A substance is said to be pozzolanic when, while not necessarily cementitious by itself, it possesses constituents which will combine with hydrated lime at ordinary temperatures in the presence of moisture to form stable insoluble compounds of cementitious value. When added to lime mortar in addition to or in partial substitution for the sand it will impart to it hydraulic properties and greater strength. When added to concretes and mortars made with Portland and similar cements, by the consequent removal of the free lime, it will impart to the material an increased degree of resistance to various agencies which are normally liable to cause disintegration.

Strength of Mortars

In the Historical Section, I, it was pointed out how the ancient Romans were able to produce magnificent long-enduring and water-tight structures with lime mortar by the use of the *volcanic sand* from the neighbourhood of Mount Vesuvius. In the next Section VIII, " Mortar sand Concrete ", it will be shown how all disinterested authorities are unanimous in admitting the limited degree of strength developed in ordinary Class A fat lime and sand mortars when in thick masses, or in the interior of solid walls and masonry pillars, &c., and the excessively slow nature of the process of carbonation to which is to be attributed any hardening after the first set and desiccation of the mass.

Roman Pozzolana

In connexion with the Classification of Limes, Section III, mention was made of the hydraulic limes, which have a definite setting power independent of absorption of atmospheric carbon dioxide, and hence are able to harden under water. These were limes containing argillaceous matter, *i.e.*, hydrated aluminium silicate, and therefore combined silica. The ancient Greeks and Romans had found (and the fact was utilised by Smeaton in the historic instance of the building of the Eddystone Lighthouse) that the clayey or siliceous matters do not need to be already present as an ingredient in the limestone before burning, but can be added subsequently, provided only that these additions are "pozzolanic." Then the mortar resulting from the admixture of pozzolanic material with the slaked lime and sand will show marked hydraulic properties and a development of considerable strength on setting independent of any slow and uncertain process of carbonation, even though a fat, non-hydraulic lime had been used. In spite of the competition of artificial Portland cement, such pozzolanic lime-cements are still prepared, and have been used on large scale outside of Great Britain (particularly in districts where a good pozzolana is available), in massive works where the ultimate degree of mechanical strength is not called for, and where cost of production is an important item. Since suitable pozzolanic material can be made available in England, there is no reason why its valuable properties should not be utilised in connexion with ordinary brick-work and masonry.

Classification

Pozzolanas can be roughly divided into (A) natural pozzolanas, (B) artificial (*i.e.*, artificially burnt) pozzolanas. (A) The ***natural pozzolanas*** are principally clayey or siliceous matters of volcanic origin, which have experienced in their formation a heat treatment analogous to that which we give to the artificial substitutes. The characteristic example is the historic "pozzolana" (puzzolana, pozzuolana) named after the town of Pozzuoli, near Naples and Mount Vesuvius, in Italy, where it abounds. This has been worked for centuries by small open pits, the material being simply screened and ground without any particular control, being therefore of variable quality. A similar material is found in areas in the south-east of France, in the Azores, and in the Canary Islands ("tosca"), whence it is exported. Some of the French pozzolana and the "tetin" from the Azores has a large percentage of iron (ferric oxide) present, and is reddish in colour. Recently, experiments have been tried with Japanese volcanic ash.

An important pozzolanic material, which has become better known of late years in Germany, is "trass," an ancient volcanic mud occurring along the Rhine, particularly at Andernach, near

Coblenz. This is now a yellow-grey rock of a pumice-like nature, of chemical composition very closely resembling that of Italian pozzolana, but in general with a rather lower iron content. The "Santorin" ash, from one of the Grecian isles, has a high content of silica and but little iron present. The composition of most of these natural pozzolanas shows about 50 per cent. of silica and around 30 per cent. of alumina and iron, together with small quantities of lime and smaller of magnesia, as main constituents. Average detail analyses are given in the Appendix to this Section.

Feebly pozzolanic materials are of fairly common occurrence: "arênes," "psammites," and sands from the decay of igneous rocks, particularly basic rocks such as basalt; and from schists. These have not had the same thermal treatment as the genuine pozzolanas, but give when used in place of ordinary sands an ultimate strength considerably above that of a fat-lime-sand mortar. The action is apparently the same as with the volcanic pozzolanas, their activity being possibly due to the presence of notable proportions of "soluble" silica, which can unite with the lime to form calcium silicates.

(B) *Artificial pozzolanic materials* include pounded bricks and tiles ("potsherds," of historic interest); burnt clay; and granulated blast-furnace slag. The last is used in the slag-cement industry, where a liquid slag is finely granulated by bringing it into contact with water or steam, then dried and ground

Burnt clay, burnt to a light terra-cotta colour, has been used effectively as an artificial pozzolana; *e.g.*, in the Asyût Nile Barrage, 1902. The common use of potsherds in early historical times has been mentioned in the first Section; some ground bricks form a convenient modern substitute for these, and if added to a fat lime mortar will have similar effect on the setting powers, strength, and hydraulic qualities of the lime. Burnt clay, "surkhi," is commonly used for this purpose in India.

Other artificial pozzolanic materials which have been tried or suggested for use are spent oil shales, residues from aluminium manufacture, "Si-Stoff," burnt "gaize," lightly burnt pozzolanic earths, and clinker. An interesting application of pozzolana is that of addition to a Portland cement in order to fix the free lime normally present in set Portland cement, for fire-proofing purposes; baked clay burnt at a temperature around 850° C. proving effective in this capacity when finely ground.

Practical Test for Pozzolanic Properties

A simple practical test to determine whether a given sample of burnt clay or other material possesses pozzolanic properties, and (to some extent) to obtain a measure of its relative activity as compared with other samples, is that utilised in connexion with the investigation on the fire-proofing of Portland cement by the addition of pozzolanas to combine with the free lime, already referred to. This consists in adding a small quantity of the material, finely powdered, to rather less than its weight of slaked lime, in a tube or small narrow bottle, and covering it to a depth of a couple of inches with pure water ; shaking up every 12 hours for a week ; and then observing the appearance of the tube (shortly after shaking) and the bulk of the sediment, as compared with a fresh mixture or with another pozzolana similarly treated. The actual quantities recommended are 0·5 gm. powdered pozzolana, 0·3 gm. slaked lime, and 20 c.c. distilled water, in medium-sized test-tubes.

The interaction between the solution of lime in water and the pozzolana will give rise to hydrated calcium alumino-silicates of complex or indefinite chemical structure ; these hydrates will be formed in the colloidal condition, and are much more bulky than the clay or lime, with the result that after seven days the tube will contain an increased volume of solid matter as measured by the height it extends up the tube, also the rate of settlement, owing to its flocculent nature, will be slower ; the activity of the pozzolana can therefore be judged by inspecting the tubes at different times after shaking. In comparison, the volume of the solid matter in a similar tube immediately after mixing will be much smaller, and it will settle almost immediately.

APPENDIX

Average Analysis, Natural Pozzolanas

(31 samples)

	Per cent.
Silica	42–66
Alumina	15–17
Iron oxide	4–19
Lime	3– 7
Magnesia	5– 7
Water	4– 9

From ECKEL, *Cements, Limes and Plasters.*

References

BURCHARTZ, H., *Mitt. Materialprüfungsamt*, 1912, **3**, 133. (Pozzolanas.)

BURCHARTZ, H., *Mitt. Materialprüfungsamt*, 1921, **39**, 240. (Trass, etc.)

BURR, *Ancient and Modern Engineering*. New York (1902). (Roman Concrete.)

ECKEL, E. C., *Cements, Limes and Plasters*, loc. cit, Part VII. (An extensive bibliography of slag cements is included.)

FEICHTINGER, G., *Die Chemische Technologie der Mörtelmaterialien*. Braunschweig, 1885. (Chapt. II, 2, a.)

FERET, R., *Rev. mat. constr. trav. publ.*, 1925, **192** ; 1926, **204** (Pozzolanas : Proportions and effect of particle size, etc.).

JOHNSON, *Materials of Construction*, loc. cit. (p. 357).

KIEPENHAUER, L., *Kalk und Moertel, Köln*, 1907 (especially p. 356).

KILLIG, *Bauing*, 1925, **23**, 1008. (Swedish Oil-Shale Residues.)

MALQUORI, G., *Gazz. chim. ital.*, 1926, **56** (1), 42. (Thermal behaviour of pozzolanas.)

MITCHELL, *Building Construction, Advanced Course*, loc. cit. (p. 13).

RITZMANN, F., *Dissertation. Heidelberg* (1907), Untersuchungen ueber Trass-, Kalk-, Sandmoertel.

RIVINGTON, *Notes on Building Construction*, loc. cit. (Part III, p. 201).

ROORKEE PAPERS, Thomason Civil Eng. Coll., India, loc. cit., Vol. I, Sect. I, paras. 138, 139, 140, 142. (Surkhi and Hydraulic Mortar.)

SHOJI, *Jap. Ceram. Assoc. J.*, 1925, **33**, 41.

SHOJI, *Ceram. Soc. Trans.*, 1924–25, **24** (4), 78.

SHOJI, *Soc. Chem. Ind. J.*, 1925, **44**, 851 B. (Abstract). (Japanese Volcanic Ash.)

STRADLING, R. E., and BRADY, F. L., Department Scientific and Industrial Research, *Building Research Spec. Rep. No.* 8. (1926.) Fire Resistant Construction. (Pozzolanic Additions to Portland Cement.)

—— *Tonind. Ztg.*, 1922, **46**, 90–2. Official German Specifications for Trass.

—— Ibid., Numerous References on Trass, particularly in 1923–26.

VITRUVIUS, *Architecture* (loc. cit.). Book II, Chapter VI (of Pozzolana).

SECTION VIII.—MORTARS AND CONCRETE

Definition

MORTAR is defined as any material used in a plastic state which can be trowelled, and becomes hard in place, and which is utilised for bedding and jointing. The word "mortar" is thus used without regard to the composition of the material, but simply defining its use as a bonding material, and to distinguish it from "stuccos" and "plasters."

In this Report we shall necessarily deal only with *lime* mortars, made with limes of any of the Classes A, B or C, and sand, with or without pozzolanic additions; and lime concretes. The latter are evidently only a special case of lime mortar, wherein the cementing material unites the particles of an aggregate consisting largely of gravel or crushed stone, &c., of a size much larger than the particles of sand which form the whole aggregate in ordinary mortar, in place of uniting bricks, ashlar stone blocks or rubble blocks. What is said here with regard to the setting process and strength of lime mortars applies with equal force to lime concretes.

Setting of Mortars (Pure or Fat Lime Mortars)

Reference has already been made in preceding Sections to the universal acceptance of the principle that fat lime mortars have but little power of *setting*, other than by simple desiccation accompanied by the development of a small degree of mechanical strength. Vitruvius upholds the practice of adding to lime mortar pozzolanic materials such as ground potsherds and Vesuvian ash. Vicat says of fat limes: "Their use ought for ever to be prohibited, at least in works of any importance." Sir Charles Pasley states of fat lime mortar: "When wet it is a pulp or paste, and when dry it is a little better than dust." Allen says: "Pure lime, in itself, has neither power of setting nor ultimate strength, and it is only when combined with clay and burnt that it is of value as a building material. . . . It is only fit for plastering and other sanitary purposes." Rivington speaks of the "superficial carbonation, less than ½-in. thick, the interior being a soft pulp or friable powder." Mitchell states: "Pure lime mortar built in thick walls never hardens or sets, but crumbles into a friable powder." It is stated in the Cyclopedia of Civil Engineering: "The hardening of common lime mortar is due to the formation of a carbonate of lime by the absorption from the atmosphere of carbon dioxide. This will penetrate for a considerable depth in course of time; but instances are common in which masonry has been torn down after having been erected many years, and the lime mortar in the interior has been found still soft and unset, since it was hermetically cut off from the carbon dioxide of the atmosphere. For the same reason, common lime mortar will not harden under water, and therefore it is utterly useless to employ it for work under water or for large masses of masonry." A German writer, Nuesbaum, condemns lime mortar for use below ground level, or except for light loads and where there are no tensile stresses, stating that the useful penetration of carbon dioxide is but $\frac{2}{5}$-in. in horizontal joints and $\frac{1}{2}$-in. in vertical joints.

In 1911 Dibdin reported that free lime was present in most of the samples of mortar taken from St. Paul's Cathedral, as much as 20 per cent. being found in one specimen taken from the core of a main pier; and several writers mention the presence of free lime, or the friable nature of the mortar in other old buildings. In connexion with plaster, it has been noted that, even in comparatively thin layers, fat lime mortars will show free lime after the passage of thousands of years.

Fat and Hydraulic Limes

It is evident, therefore, that the process of carbonation, to which is usually ascribed the hardening of lime mortars, cannot be depended upon to develop any considerable strength in thick brick walls or masonry when fat, high-calcium Class A limes are used without any pozzolanic addition; and it is clear why the Class C hydraulic limes, such as the blue lias lime, have been so much valued for exacting work, since they are to some extent independent of carbonation for the development of their strength. Reference should be made again at this point to the fact that the colour of an alleged hydraulic or "Blue Lias" lime does not necessarily imply hydraulic properties; and that there are poor, lean, grey Class B limes with little or no hydraulic powers and working qualities inferior to those of the weak Class A limes. There is evidence, however, that *magnesian* (dolomitic) limes give mortars that eventually develop greater strength than those made of high-calcium limes. A clear statement of the fundamental difference between a fat lime-sand mortar and one in which hydraulic properties have been developed by the addition of a pozzolanic substance, such as clinker, is given—in connexion with fire-resisting properties of mortars—in the Building Research Special Report No. 8 referred to in the Bibliography at the end of Section VII.

Strength of Lime Mortars

Much that has been said in Section VI, "Plaster," as to shrinkage, sand-carrying capacity, and the processes involved in the slaking and tempering of limes, in as much as these factors affect ultimate strength, will apply also to lime mortars. But with mortars the matter of ultimate compressive strength is of importance; and occasionally the degree of adhesion and tensile strength will come into question. In comparison with the vast amount of work carried out in connexion with Portland cements, there are comparatively few recent data of a reliable order for lime mortars. The reason for this is not hard to seek ; the compressive strength of even a poor lime mortar is more than adequate for any ordinary load in *small* buildings, and the adhesion and tensile strength are rarely likely to be tested in connexion with *ordinary* building construction. In any critical situations it has become customary to substitute a Portland cement of known and fairly narrowly specified properties. Reference has already been made to the common ancient practice of dispensing entirely with mortar in large stone or other constructions, or at most using it to make good defects and crevices ; and to the evidence that many mediæval buildings owe their strength rather to massive design than to hardened mortar. In tests of the strength of brick piers under compression, occasionally it is found that a greater ultimate strength will be developed when the bricks are carefully imbedded in sand alone, without any admixture, than when built up with lime mortar ; though in the former case no tensile strength and but little shearing strength will be shown.

Practical Limits

Estimates of the compressive load imposed on the lowest courses in ordinary small house construction show that the maximum attained is well below the 100 to 150 lb. per sq. in. maximum which has been suggested for a safe working stress for lime-mortar brick construction ; and this figure is again very much below the compressive strength exhibited in tests to destruction of brick piers, &c. There are cases on record of lime mortar-brick construction (classification of lime not specified) standing far greater loads than those implied here ; *e.g.*, in a 400-foot chimney in Glasgow.

Strength of Hydraulic Lime Mortars

With hydraulic lime mortars, appreciable tensile strength can be exhibited ; figures ranging from 46 lb. to nearly 300 lb. per sq. in. are given, for mortars of different degrees of hydraulicity, and for 1 year old samples. On the other hand, in connexion with fat lime mortars it is well known that the samples may fall apart in the attempt to fix them in the machine to determine tensile strength or adhesion, after " setting " for a considerable time. A comprehensive series of tests with high-calcium, lean, and hydraulic limes, in conjunction with four different types of sand, carried out in 1911 by Dibdin for the Royal Institute of British Architects, brought out clearly the difference between the various types of lime, in setting power and ultimate strength developed. Extensive data for Continental hydraulic limes are also given by Burchartz (1912).

Sand in Mortars

In addition to what has been dealt with in connexion with sand in plasters, the series of tests by Dibdin emphasised the importance of the sand containing particles of varying sizes as well as the more obvious matter of right proportioning of the sand in mortars. The hydraulic and lean limes, containing appreciable quantities of other materials than calcium and magnesium hydrates, in a fine state of division, can tolerate without serious loss of ultimate strength very much less sand than the Class A, fat, high-calcium limes ; particularly when the sand itself contains a large proportion of the finer sizes. A moderately coarse sand, which has a uniform distribution of particles between the maximum and minimum sizes, appears best ; if this is also of a pozzolanic character, such as ground brick or some types of clinker or granulated slag, an excellent mortar results. The actual proportions of sand for maximum strength approach 2 to 1 of lime for hydraulic mortars, and for Class A lime mortars the results of Dibdin's tests appear to indicate a maximum strength with rather less than the customary 3 : 1 proportions, when sand with particles of not too uniform a size is used. Compressive strength, tensile strength, and adhesion appear to vary approximately proportionately when different mortars are compared.

Sharp pit sand has been preferred from the time of Vitruvius to the more rounded river sand. Sea sand should be avoided, for the reasons indicated in connexion with plaster. Apart from pozzolanic properties, finely crushed rock will often replace the natural sand quite satisfactorily ; cleanliness, and a uniform distribution of particle size, together with an average size not too fine, appear to be of greater importance than actual form of particle.

Plasticity of Mortars

The matter of plasticity is far less exacting in the case of mortar, with its relatively simple technique in building operations, than in the case of plaster ; hence the greater plasticity of Class A limes can be sacrificed for the sake of ultimate strength, and the less plastic Class C limes are preferable, when a pozzolana is not also used. A greater plasticity implies here, as in plaster, a greater *practical* sand-carrying capacity when ultimate strength is not a predominating factor, the more plastic lime allowing of a larger proportion of sand without losing its free-working properties ; though the strength is diminished, as explained above, if the sand is unduly increased.

Storage and " Knocking-up " of Hydraulic Mortars

It has been pointed out in connexion with plasters, which are made preferably with plastic Class A limes, that a long period of storage of the lime putty, and subsequent working up of the mass before use (" tempering "), were not only harmless but actually very highly beneficial, in that the plasticity or easy spreading quality of the plaster is promoted to the full extent. In fact, as mentioned before, a lime for plaster cannot be tempered too long. However, it is otherwise with hydraulic limes ; an ample period for completing the somewhat sluggish slaking process here is desirable, but long storage of mixed hydraulic mortar, with subsequent working up of a material which may already be partly set, is evidently accompanied by serious danger of reducing its final strength when completely set, and should be avoided. The time-limit for storage of hydraulic mortars is given in the text-books as some nine days ; the actual maximum period permissible will obviously depend on the degree of hydraulicity of the particular mortar, and its speed of setting, and for good work should be previously determined.

Slaking of Limes for Mortar

The proper slaking of Class A limes has been discussed already in connexion with plasters. When used for mortars, the same care should be taken to avoid over-heating the lime during the slaking process, and to remove unslaked lumps of overburnt or underburnt lime after slaking, as well as to allow a reasonable time for the process to complete itself ; but the long period of tempering advisable in connexion with lime-putty to be used in plaster is not called for in connexion with mortar.

Magnesian limes slake slowly ; in fact a proportion of the magnesium oxide present may completely escape hydration under ordinary working conditions, particularly if burnt at a high temperature. Accordingly there is little danger of magnesium limes overheating during slaking, and they can be safely treated in a similar manner to hydraulic limes. With lean Class B limes, it is best to conserve the heat produced in the slaking process ; therefore after the water is added to the lime, the latter should be covered up with a layer of sand and left for some time for the process to come to an end. Any large unslaked particles of overburnt or underburnt lime should then be removed ; overburnt lime may slowly slake after the mortar is in position, and (just as in " lime-pops " in plaster) may cause damage by its consequent expansion. For this reason, some specifications require the slaked lime to be run through, say, a No. 8 or No. 10 sieve or screen before using. As with plaster, the use of finely divided ready hydrated lime greatly simplifies the process of mortar mixing.

Eminently hydraulic limes may show so little activity in the slaking process that it is better to have them in a finely ground condition before commencing the slaking ; and to allow plenty of time for this operation.

Lime Concrete

According to some authorities, in certain classical buildings which survive to this time, a pozzolanic lime-mortar concrete was used, poured in the modern fashion in temporary moulds, or more often permanently faced and reinforced by bricks, ashlar blocks, or triangular tiles arranged horizontally in layers, the concrete having been rammed in place between these. Vitruvius speaks thus : " tumble in the rubble between them (*i.e.*, the brick or masonry walls) " ; " filled up with small rubble run with mortar " ; and Pliny describes the construction of cisterns with what appears to have been a rough rubble, or concrete placed in position piecemeal rather than poured. Desch quotes the vaults of the Thermae and the Basilica of Constantine as examples of genuine *poured* lime concrete construction in Roman times. The Chinese used at an early date a lime mortar plus aggregate in military works ; but here the materials were put in position dry, and the water added last.

Except in these few favourable classical examples, or where by chance a hydraulic lime had been used, such rough rubble work in the filling of large piers or ashlar-faced walls, &c., usually exhibits, after the passage of hundreds of years, a lack of cohesion and mechanical strength which emphasises once more the point as to the necessity of using either a hydraulic lime, or else some pozzolanic material in conjunction with a high-calcium lime. With the latter materials, a good strong hydraulic mortar and concrete will result, though more slow in setting than the artificial (Portland, &c.) cements or " Roman " cements. However, as with the latter and with other " natural " cements, lack of uniformity and absence heretofore of any exact standards of quality greatly handicap these natural lime products in competition with the artificial and closely controlled product. The latter is therefore likely to continue to be preferred for exacting work in concrete, particularly as the setting time can be adjusted in modern sintered or fluxed cements to suit particular requirements, whereas, as at present developed, hydraulic or pozzolanic lime concrete is necessarily slow setting.

References

ALLEN, *Practical Building Construction*, 1903, p. 364.

—— *Amer. Soc. Testing Materials Proc.*, 1924, **24**. Tentative Specifications and Definitions, C5, 35, 51, 11, 58T.

—— *Amer. Soc. Testing Materials Proc.*, 1924, **24**, p. 404. (Time of Setting of Mortar.)

BURCHARTZ, H., *Luftkalke und Luftkalk-moertel.* Berlin (1908).

BURCHARTZ, H., *Hydraulische Kalke.* Berlin (1912).

BURNELL, G. R., *Elementary Treatise on Limes, Cements, Mortars, etc.* (1857).

BURR, *Ancient and Modern Engineering*, loc. cit.

—— *Cyclopedia of Civil Engineering* (1913), (Vol. IV, pp. 14, 18.)

DESCH, C. H., *Chemistry and Testing of Cements*, loc. cit.

DIBDIN, *Roy. Inst. Brit. Arch. J.*, 1922 (Aug. 19). (Ancient Mortars.)

DIBDIN, *Roy. Inst. Brit. Arch. Report*, 1911. (Tests of Limes and Sands.)

ECKEL, E. C., *Cements, Limes and Plasters*, loc. cit.

EMLEY, *Amer. Ceram. Soc., Trans.*, 1914, **16**, 11. (Penetration of Carbon Dioxide.)

EMLEY and YOUNG, *Amer. Soc. Testing Materials*, 1914, **14**, 339. (Magnesian Lime Mortars.)

—— *Engineering News*, 1904 (June 9), **51**, 543. (Tensile Strength.)

Feret, *Les materiaux de construction*, 1904, **1**, 139.

Knibbs, N. V. S., *Lime and Magnesia*, loc. cit. (Chapter II, p. 248.)

Mitchell, C. F., *Building Construction, Advanced Course*, loc. cit. (Chapter I.)

National Lime Assoc. (U.S.) *Pamphlets*, loc. cit. Construction, Chapter II, III and IV.

Neumann, E., *Kalkbeton*. Berlin (1923).

Nuesbaum, *Zement*, 1924 (6). (Penetration of Carbon Dioxide.)

Rivington, *Notes on Building Construction*, loc. cit. Part III, Chapter III.

—— Specification (1926), loc. cit. (*Bricklayer*. p. 72.)

Stradling, R. E., and Brady, F. L., Department Scientific and Industria Research, *Building Research Special Report No.* 8, loc. cit.

U.S. Bureau of Standards, Circ. No. 30. Lime, its Properties and Uses, loc. cit. (Sects. III, V).

SECTION IX.—SPECIFICATIONS

SPECIFICATIONS of building materials are of two types : (*a*) those for use between purchaser and vendor, the clauses of which are intended to ensure that the whole of the material bought shall be of the required quality and up to sample, (*b*) those for use between architect and contractor, the clauses of which are intended to ensure that only suitable materials are used and that the finished work is of the required quality.

Accordingly, there are given here (*a*) tentative Specifications for *Quicklime* for use in building construction, Class A, high magnesian Class A, and Class B ; *Hydrated Lime*, for use in building construction, Classes A and B ; and *Hydraulic Lime*, Classes C 2 and C 3 ; and (*b*), Specifications for *Lime Putty* and *Lime Mortar*, and materials used in preparing these.

(*a*) SPECIFICATIONS FOR USE BETWEEN PURCHASER AND VENDOR

Tentative Standard Specification for Quicklime for Use in Building Construction : Plaster and Mortar

CONTENTS

Clause 1. Description.
,, 2. Composition.
,, 3. Residue.
,, 4. Samples for Testing.
,, 5. Sampling.
,, 6. Facilities for Sampling and Identifying.
,, 7. Cost of Tests, Analyses and Samples.
,, 8. Non-Compliance with Tests.

(N.B.—Clauses 6, 7 and 8 are of a purely formal character, corresponding precisely with Clauses 2, 5, 6 and 14, in the British Standard Specification for Portland Cement, B.E.S.A., 1925, and hence are not given at length here.)

Description

1. The material shall consist essentially of calcium oxide with or without a smaller proportion of magnesium oxide, formed by calcining a natural rock at such a temperature that it will slake readily when brought into contact with water. It shall be one of the following grades as specified by the purchaser :—

Quicklime, Class A, for plaster, fine and coarse stuff.
High Magnesian Quicklime, Class A, for plaster, fine and coarse stuff.
**Quicklime, Class B*, for plaster, coarse stuff.

Either grade, Class A or Class B, shall be suitable for use in mortar when gauged with an approved pozzolanic material.

* Class B: A high magnesian Class B quicklime is not differentiated.

Summary of Tests

Test	Quicklime			Hydrated Lime		Hydraulic Lime		
	Class A	High Magnesian Class A	Class B	Class A	Class B	Class C 1	Class C 2	Class C 3
Composition	Not less than 95% CaO & MgO* Not more than 10% CO_2 — * On ignited sample.	Not less than 95% CaO & MgO* Not more than 10% CO_2 Over 20% MgO*	Not less than 85% CaO & MgO* Not more than 10% CO_2 —	Not less than 95% CaO & MgO* Not more than 5% CO_2 —	Not less than 85% CaO & MgO* Not more than 5% CO_2 —	Not less than 60% CaO & MgO* Not more than 10% CO_2 Not more than 15% SiO_2 & Al_2O_3*	Not less than 60% CaO & MgO* Not more than 10% CO_2 Not more than 25% Not less than 15% SiO_2 & Al_2O_3*	Not less than 60% CaO & MgO* Not more than 10% CO_2 Not more than 35% Not less than 25% SiO & Al_2O_3*
Fineness and Residue	Not over 5% when washed through No. 20 sieve			All through No. 20 sieve Not more than 10% on No. 200 sieve Not more than 1% on No. 50 sieve	All through No. 20 sieve Not more than 15% on No. 200 sieve	*Either* not over 5% when washed through No. 20 sieve *Or* ground to pass No. 20 sieve, and not over 15% retained on No. 200 sieve		
Soundness	—	—	—	Accelerated steaming test on ¼″ sand-lime pat, 2 days old; 5 hours in steam	—	—		
Tensile Strength, 3 : 1 (wt.), Standard Sand-Lime Briquettes	—	—	—	—	—	—	At 28 days: Stored 72 hrs. in moist air, balance of 25 days in water: not less than 75 lb./sq. in.	

Composition

2. Class A: Not less than 95 per cent. by weight of the ignited sample shall consist of calcium and magnesium oxides; the amount of carbon dioxide present shall not exceed 10 per cent. of the weight of the material as received.

If the amount of the magnesium oxide present in a Class A quicklime exceeds 20 per cent. of the weight of the ignited sample it shall be termed a High Magnesian Class A quicklime.

*Class B: Not less than 85 per cent. by weight of the ignited sample shall consist of calcium and magnesium oxides; the amount of carbon dioxide present shall not exceed 10 per cent. of the weight of the material as received.

Residue

3. When broken to pass a 1-in. ring, slaked carefully by immersion in water and then allowed to stand covered by water for 24 hours, the residue left on a standard Institute of Mining and Metallurgy No. 20 sieve after washing the material through with a jet of water shall not exceed 5 per cent. of the weight of the lime sample as received, after drying the said residue to constant weight at 110° C. The washing operation shall not occupy more than 30 minutes, and the lime shall not be rubbed through the sieve. The amount of lime taken for this test shall not be less than 2 pounds.

Samples for Testing

4. Three samples of approximately equal weight, and each sample in no case less than five pounds in weight, shall be taken, and enclosed immediately in air-tight containers. One sample shall be assigned to the purchaser, one to the vendor, and the third retained for submission to a referee in case that an analysis or test be subsequently questioned.

Sampling

5. When the lime is loose in piles or trucks, approximately equal portions shall be selected from at least twelve different positions so as to be representative of every part of the material, and placed in a pile on a clean floor. Not less than 50 pounds in all shall be so taken. The material in the composite sample shall be thoroughly mixed and if necessary any larger lumps broken to pass a 1-in. ring; then quartered down, with repeated mixing of all parts, until only sufficient remains to provide the three samples. It shall then be divided into three approximately equal parts, and immediately placed in the three air-tight sample containers.

With bulk lime in movement, on conveyers or in barrows, &c., samples shall be taken at regular intervals during the whole period of movement, so that the same proportion is obtained,

* Class B: A high magnesian Class B quicklime is not differentiated.

and a total weight of sample of not less than 50 pounds. This shall then be treated in the same manner as indicated in the first part of this clause to obtain the final samples.

With lime in bags, barrels, or other packages, samples shall be taken from not less than twelve different bags, barrels, or other packages, or where there is a less number than twelve different bags, barrels, or other packages, then from each bag, barrel, or other package. Where the lime is in lump form, sufficient material shall be taken from each bag, barrel, or other package sampled to produce a total combined sample of not less than 50 pounds weight, this material being taken so as to be as far as possible representative of the contents of the bag, barrel or other package, including both lump and powder.

Where the material is substantially all in powder form, a sampling tube shall be used of approximately 1-in. diameter, and the samples shall be taken with this sampling tube so as to be representative of the whole contents of the bag, barrel, or other package sampled throughout its depth or thickness, so as to produce a total combined sample of not less than 50 pounds weight. Whether in lump or powder form, the combined sample shall then be treated in the same manner as indicated in the first part of this clause to obtain the final samples.

The process of sampling shall be carried out as quickly as possible so as to avoid unnecessary exposure of the material to the air; no samples shall be taken from broken packages.

Facilities for Sampling and Identifying

6. (Formal: *vide* British Standard Specification for Portland Cement, B.E.S.A., 1925, Clauses 2 and 5.)

Cost of Tests, Analyses and Samples

7. (Formal: *vide* British Standard Specification for Portland Cement, B.E.S.A., 1925, Clause 6.)

Non-Compliance with Tests

8. (Formal: *vide* British Standard Specification for Portland Cement, B.E.S.A., 1925, Clause 14.)

Notes

Attention should be drawn again to the difference between the different types of limes, Classes A, B, C 1, C 2 and C 3 (Section III, " Classification "). A specification for quicklime for plastering will necessarily differ widely from one for a moderately (C 2) or eminently (C 3) hydraulic lime suitable for good structural mortar; whilst by proper pozzolanic additions the non-hydraulic limes can generally be rendered suitable for use in structures under heavy loads, or for work below the ground-level.

Plasticity

It would be very desirable to be able to specify precisely the degree of plasticity demanded in the different grades of building lime, particularly in lime for plaster-work; but (as indicated in Section VI, " Plaster," paragraph " Plasticity ") on account of the indefiniteness of this complex property we do not possess, at present, any fully satisfactory method of measuring the plasticity of lime plaster or mortar, unless indeed the Emley plasticimeter be adopted officially for this purpose.

The best that can be done at present is to specify the chemical composition and mode of slaking and maturing the putty for Class A limes, so as to obtain a reasonable assurance that the lime will possess a sufficient degree of plasticity in working for exacting plaster-work.

Tentative Standard Specification for Hydrated Lime for Use in Building Construction : Plaster and Mortar

CONTENTS

(N.B.—Clauses 7, 8 and 9 are of purely formal character, corresponding precisely with Clauses 2, 5, 6 and 14 in the British Standard Specification for Portland Cement, B.E.S.A., 1925, and hence are not given at length here.)

Description

1. The material shall be a fine dry powder, produced by treating quicklime with just sufficient water to satisfy its chemical affinity, and shall consist essentially of calcium hydroxide with or without a smaller proportion of naturally associated magnesium oxide and hydroxide. It shall be one of the following grades as specified by the purchaser :

Hydrated Lime, Class A, for plaster, fine and coarse stuff.
Hydrated Lime, Class B, for plaster, coarse stuff.

Either grade, Class A or Class B, shall be suitable for use in mortar, when gauged with an approved pozzolanic material.

Composition

2. Class A : Not less than 95 per cent. by weight of the ignited sample shall consist of calcium and magnesium oxides ; the amount of carbon dioxide present shall not exceed 5 per cent. of the weight of the material as received.

Class B : Not less than 85 per cent. by weight of the ignited sample shall consist of calcium and magnesium oxides ; the amount of carbon dioxide present shall not exceed 5 per cent. of the weight of the material as received.

Test for Fineness

3. The hydrated lime shall comply with the following conditions of fineness: 100 gm. of the hydrated lime shall be continuously sifted for a period of five minutes on an Institution of Mining and Metallurgy Standard No. 20 sieve, and shall leave no residue. Airset lumps in the samples may be broken down with the fingers, but nothing shall be rubbed on the sieve.

Class A hydrated lime shall after passing the No. 20 sieve be then continuously sifted for a period of fifteen minutes on an Institution of Mining and Metallurgy Standard No. 200 sieve and for a period of five minutes on an Institution of Mining and Metallurgy Standard No. 50 sieve and in the order of succession given below with the following results:—

(1) The residue from 100 gm. on the No. 200 sieve shall not exceed 10 gm.

(2) The residue from the same 100 gm. on the No. 50 sieve shall not exceed 1 gm.

Class B hydrated lime shall, after passing the No. 20 sieve as hereinbefore specified, be continuously sifted for a period of fifteen minutes on an Institution of Mining and Metallurgy Standard No. 200 sieve; the residue on the No. 200 sieve from 100 gm. shall not exceed 15 gm.

Test for Soundness

Class A

4. Hydrated lime, Class A, shall comply with the following test for soundness. A pat 4 in. diameter and $\frac{1}{4}$ in. thick of mortar, made with one part by weight of the lime to five parts by weight of standard sand, allowed to set in air for 24 hours and then covered with a thin coat of fresh lime paste and allowed to set another 24 hours, shall show no signs of cracking, pitting, or disintegration after immersion for five hours in saturated steam. The standard sand used shall be white Leighton Buzzard sand identical with that specified in Clause 11, Test for Tensile Strength (Cement and Sand) in the current British Standard Specification for Portland Cement; the lime and sand shall be mixed with such a proportion of water, in making the test pat, that the mixture shall be plastic.

Samples for Testing

5. (Identical with Clause 4, Specification for Quicklime.)

Sampling

6. (First two paragraphs identical with first two paragraphs of Clause 5, Sampling of Specification for Quicklime.) (Third paragraph.)

With hydrated lime in bags, barrels, or other packages, samples shall be taken from not less than twelve different bags, barrels, or other packages, or where there is a less number than twelve different bags, barrels, or other packages, then from each bag, barrel, or other package.

A sampling tube shall be used of approximately 1 in. diameter, and the samples shall be taken with this sampling tube so as to be representative of the whole contents of the bag, barrel, or other package sampled throughout its depth or thickness, so as to produce a total combined sample of not less than 50 pounds weight. This combined sample shall then be treated in the same manner as indicated in the first part of this clause to obtain the final samples.

The process of sampling shall be carried out as quickly as possible so as to avoid any unnecessary exposure of the material to the air ; no samples shall be taken from broken packages.

Facilities for Sampling and Identification

7. (Identical with Clause 6, Specification for Quicklime.)

Cost of Tests, Analyses and Samples

8. (Identical with Clause 7, Specification for Quicklime.)

Non-Compliance with Tests

9. (Identical with Clause 8, Specification for Quicklime.)

Notes

Since the purpose of hydrating the lime mechanically at the limeworks is to obtain a finely powdered product which can be stored indefinitely in bulk or in bags without serious deterioration, it is evident that an eminently hydraulic Class C 3, or moderately hydraulic Class C 2 lime is not suitable for this purpose, even if sufficiently quick in slaking to be handled satisfactorily in the hydrator ; unless the somewhat elaborate procedure of the French Grappier cement industry be adopted. Accordingly, only Class A and Class B limes are likely to come into question.

A specification for hydrated lime for use in plaster-work must ensure that the lime is free from those defects and impurities that have been found to result in pitting and unsoundness in plaster. (Compare Section VI, " Plasters," paragraphs on " Pitting," etc.). The requirements as to fineness must be rigidly defined, for this reason, as well as to ensure a high plasticity and sand-carrying capacity. *Re* Plasticity, *vide* Note in Specification for Quicklime.

The test for soundness, Clause 4, is needed in this case, and is omitted in the case of quicklime, because the purchaser expects to use hydrated lime immediately, or at latest after a soaking overnight ; whilst in the latter case the soundness of the lime putty will necessarily depend on the method and time of maturing adopted. Slow-slaking and hard-burnt particles, or impurities which may cause pitting later on, will manifest themselves in an unmistakable manner in this last test. It does not appear necessary, then, to specify a maximum for certain impurities, *e.g.*, sulphur.

The *colour* of the hydrated lime (*e.g.*, for use in a finishing coat) may form the subject of a special clause where required.

Tentative Standard Specification for Hydraulic Lime for Use in Building Construction

CONTENTS

(N.B.—Clauses 7, 8 and 9 are of purely formal character, corresponding precisely with Clauses 2, 5, 6 and 14 in the British Standard Specification for Portland Cement, B.E.S.A., 1925, and hence are not given at length here.)

Description

1. The material shall consist essentially of calcium oxide, with or without a smaller proportion of magnesium oxide, in natural association with siliceous and argillaceous matter, and formed by calcining an argillaceous limestone or chalk at a moderate temperature; which contains sufficient free lime so that it will slake on the addition of water. It shall be one of the following grades:—

Hydraulic Lime, Class C 2.
Hydraulic Lime, Class C 3.

Residue and Fineness

2. The material shall either slake and disintegrate completely when immersed in water for 24 hours, giving a residue not more than 5 per cent. of the weight of the sample as received, when tested in accordance with Clause 3, " Residue," of the Specification for Quicklime, Classes A and B; or else shall be in a finely-ground condition conforming substantially to Clause 3, " Fineness," of the Specification for Hydrated Lime, Class B.

Composition

3. The composition of the hydraulic lime shall conform to the following conditions:—

	Class C 2. Per cent.	Class C 3. Per cent.
Total calcium and magnesium oxides, by weight, of the ignited sample	Not less than 60	Not less than 60
Total silica and alumina, by weight of the ignited sample	Not less than 15 Not more than 25	Not less than 25 Not more than 35
Carbon dioxide, by weight, of the sample as received	Not more than 10	Not more than 10

Test for Tensile Strength (Hydraulic Lime and Sand)

4. The tensile strength of hydraulic lime and sand shall be ascertained from briquettes of the shape and dimensions and in the manner hereinafter specified. The breaking strength of the briquettes at 28 days after gauging shall not be less than 75 lb. per sq. in. (5·28 kg. per $cm.^2$) in the case of Class C 2 or Class C 3 hydraulic lime, this breaking strength being the average tensile breaking strength of six briquettes.

The briquettes shall be prepared in the following manner: the lime shall be slaked by sprinkling on it a sufficiency of clean water, covering up lightly, and allowing to stand for 24 hours. It shall then be gauged with standard sand in the proportion of three parts by weight of sand to one part by weight of dry lime, and with sufficient water to give the standard consistence hereinafter defined. It shall then be filled into the moulds of the form required to produce briquettes, and the whole stored in moist air for 72 hours after gauging, when the briquettes shall be removed from the moulds and immediately submerged in clean fresh water and left there until taken out for breaking.

The form and dimensions of the moulds and briquettes, the method of filling the mould, the methods of storing and of breaking the briquettes, and the standard sand used shall be identical with the corresponding form, dimensions, methods, and materials respectively specified in the current British Standard Specification for Portland Cement, Clause 11, "Test for Tensile Strength (Cement and Sand)"

The standard consistence shall be such that a 4 in. spread is recorded when a cone of the mortar is allowed to fall 20 times through ½ in. on a standard flow-table.* The cone shall be formed in a conical metal mould 3½ in. high and 2⅝ in. diameter at the lower end, 1½ in. diameter in the upper end; the "spread" shall be measured by the mean diameter of the lower end of the cone of mortar after it has settled down on the flow-table.

Samples for Testing

5. (Identical with Clause 4, Specification for Quicklime.)

Sampling

6. Identical with Clause 5, Specification for Quicklime.)

Facilities for Sampling and Identifying

7. (Identical with Clause 6, Specification for Quicklime.)

Cost of Tests, Analyses and Samples

8. (Identical with Clause 7, Specification for Quicklime.)

Non-Compliance with Tests

9. (Identical with Clause 8, Specification for Quicklime.)

* Vide *Building Research Technical Paper No. 4*

Notes

Since hydraulic lime is indicated for use more particularly in masonry and brickwork, in exacting positions and below the ground-level, or even below the water-level, a specification must demand a definite strength and definite hydraulic properties. For exacting work a certain *tensile* strength is advisable, in addition to a compressive strength which is seldom tested to the full in practical constructional work ; it is convenient therefore to bring the specification into line with the British Standard Specification for Portland Cement by demanding a tensile strength test on the standard type of briquette with standard sand in a 3 : 1 mix. At the same time, the conditions as to storage (72 hours in moist air followed by 25 days under water) give an effective test of hydraulicity. The essential control of consistence in this test is provided for here by a flow-table test ;* additional sub-clauses specify this and the mode of slaking the lime.

The specification is necessarily complicated by the fact that highly hydraulic limes may require to be supplied in a finely ground condition, as they may not slake in a reasonable space of time if in lump form.

It appears that we are not in the position, without further research,† narrowly to specify the difference between Class C 2 (moderately hydraulic) and Class C 3 (eminently hydraulic) limes on the one hand ; and between Class C 1 (feebly hydraulic) and Class B (limes with between 85 and 95 per cent. of calcium and magnesium oxides, and non-hydraulic), on the other. It is also somewhat doubtful whether the Class C 1 (feebly hydraulic limes), containing less than 15 per cent. of hydraulic constituents (silica and alumina) by chemical analysis, show any marked superiority over Class A or Class B limes in exacting constructional work without pozzolanic additions. Accordingly, Class C 1 limes are preferably treated as if belonging to Class B ; and hence are omitted from this specification altogether ; whilst Classes C 2 and C 3 are only differentiated by the positive difference in composition.

The tensile strength of 75 lb. per sq. in. proposed tentatively here represents apparently a conservative figure readily attained by a moderately hydraulic lime of good quality.

(b) Specifications for use between Architect and Contractor

Specification for Lime Putty

For Fine or Coarse Stuff, Plasterers' Work.

Contents

Description

1. Lime putty used in plasterers' work shall be a plastic material resulting from the slaking of quicklime with a sufficiency of water, or from the mixing of previously hydrated lime in a dry condition with a sufficiency of water.

* Vide *Building Research Technical Paper No.* 4.

† Work on this is now in progress at the Building Research Station.

Materials

2. Lime putty shall be made from one of the following limes, as specified by the architect in charge :—

Type I (*a*) : Quicklime, Class A, conforming to the Specification for Quicklime, Class A, for use in Building Construction : Plaster and Mortar.

Type I (*b*) : High Magnesian Quicklime, Class A, conforming to the Specification for High Magnesian Quicklime, Class A, for use in Building Construction : Plaster and Mortar.

Type I (*c*) : Hydrated Lime, Class A, conforming to the Specification for Hydrated Lime, Class A, for use in Building Construction : Plaster and Mortar.

Type II (*a*) : Quicklime, Class B, conforming to the Specification for Quicklime, Class B, for use in Building Construction : Plaster and Mortar.

Type II (*b*) : Hydrated Lime, Class B, conforming to the Specification for Hydrated Lime, Class B, for use in Building Construction : Plaster and Mortar.

Types I (*a*), I (*b*), I (*c*) may be used for fine stuff or coarse stuff ; Types II (*a*) and II (*b*) shall be used for coarse stuff only.

Slaking

3. The method of preparation of the lime-putty shall depend on the lime used, and shall conform with the methods specified herein for the corresponding types :—

Type I (*a*) : the putty shall be prepared from Class A quicklime by careful slaking by immersion, excessive rise of temperature during the slacking process or " drowning " by the use of excess water being avoided ; when the action of slaking has ceased and the putty is cold, it shall be run off through a sieve of 1/16th in. mesh, lightly covered, and stored for at least two weeks before use.

Type I (*b*) and Type II (*a*) : the putty shall be prepared from high magnesian quicklime, Class A, or quicklime, Class B, respectively, by sprinkling with the requisite quantity of water, and covering up lightly to retain the heat and moisture for 24 hours ; the material shall then be thoroughly stirred, more water being added if necessary, and run off through a sieve of 1/16th in. mesh ; then lightly covered and stored for at least two weeks before use.

Types I (*c*) and II (*b*) : the putty shall be prepared from hydrated lime, Class A or Class B respectively, by thoroughly mixing with the requisite quantity of water and storing for 24 hours before use, lightly covered.

Additions

The architect in charge may specify the addition of definite substances to the lime putty, *e.g.*, Portland cement, plaster of paris, proprietary rapid-setting cements, &c.

Sand and hair (or fibre) may be added, for coarse stuff, at any time after the screening of the putty in Types I (*a*), I (*b*), and II (*a*), or the mixing of the hydrate with water in Types I (*c*) and II (*b*).

All sand shall be hard, clean, free from vegetable or animal matter, salt, alkali, or more than a trace of iron pyrites. If demanded by the architect in charge, the sand shall be washed before use. The sand shall all pass an I.M.M. Standard No. 5 sieve, and not more than 10 per cent. by weight shall pass an I.M.M. Standard No. 80 sieve; the proportions of different particle-sizes shall be well distributed between these limits. The sand shall be pit sand, or river sand taken above the tidal zone, and not sea sand or sand from brackish waters; alternatively, approved crushed hard rock, clean hard burnt old brick, or a slag which is substantially free from sulphides, properly screened so as to conform substantially with the fineness requirements for pit or river sand, may be used. Where a mortar-mill is in use, and approved materials in lieu of sand are crushed in the mill in the process of mixing the coarse stuff, these shall be crushed so that the whole of the final product will pass through an I.M.M. Standard No. 5 sieve; the hair or fibre shall not be added until the crushing is completed.

Clear, fresh water only shall be used in mixing the materials, free from any unusual proportions of dissolved salts or gases. Sea water or brackish water shall not be used.

Hair used in coarse stuff shall be clean, sound ox or goat hair; it shall be free from grease, and shall be carefully beaten to separate any lumps before use. A substantial majority of the fibres shall be between ½ in. and 2 in. in length.

Flexible vegetable fibre, such as manilla or sisal, which otherwise conforms to this specification, may be substituted for hair. The hair or fibre shall be distributed evenly throughout the mixed plaster, in the proportion of 1 lb. of hair for every 3 cu. ft. of coarse stuff (or otherwise as directed by the architect in charge, *e.g.*, ½ lb. per cu. ft.), and not introduced until the heat of slaking of the lime is wholly dissipated.

Notes

The method of slaking quicklimes, and a minimum period of tempering for the putty, must be specified carefully in order to avoid the danger of the presence of hard burnt or other unslaked particles (above a size corresponding approximately to a No. 50 sieve) in the plaster when applied. As shown in Section VI, " Lime Plasters," there is a danger of pitting and unsoundness developing in such a case. With fat Class A quicklimes there is also a danger of burning in slaking, unless special precautions are taken. With hydrated limes conforming to the specifications named there is little danger of unsoundness, with but a brief period of tempering.

In order to avoid the possibility of pitting, efflorescence, or discoloration, it is essential that the sand used be clean and free from pyrites and similar iron-bearing minerals ; whilst for mechanical strength and smooth working there is required a proper grading of fine and coarse particles This appears to be of greater importance for sound work than the shape of the particles, whether sharp or smooth. With this limitation, either natural sand or crushed material, silica, felspar, calcite, marble, dolomite, etc., properly screened, is suitable ; or hard artificial materials such as crushed brick or slag of suitable chemical composition. Sea sand, or sand from tidal waters in estuaries, must be ruled out, on account of the danger of efflorescence due to its salt content.

If the water used is clean and fresh, the small proportions of ordinary impurities present in normally available supplies are without harmful effect, but sea water or brackish mineral-containing water must be avoided, if only on account of the dangers of efflorescence and dampness.

Specification for Lime Mortar

CONTENTS

Summary of Tests and Requirements

(*a*) *Either* Class C 2 or C 3 hydraulic lime to be used. *Or* Class A, B, or C 1 with pozzolanic additions such that tensile strength of 50 lb. per sq. in. is developed in 1 : 3 standard sand-lime-pozzolana briquettes at 28 days; stored 72 hours in moist air and balance of time in water.

(*b*) All particles in finished mortar to pass No. 5 sieve.

(*c*) Sand proportions 1 : 1 in load-bearing structures below ground or water-level; 1 : 2 above ground ; 1 : 3 in less exacting works.

(*d*) Hydraulic mortars to be used within 10 days, pozzolanic within 24 hours.

Specification for Mortar. Structural Work.

Clauses.

Description

1. Lime mortar used in structural work (*i.e.*, exclusive of plaster, stucco, and parging or pargetting of flues) shall consist of a plastic mixture of lime putty or hydrated lime, sand, and water, with or without the addition of finely divided pozzolanic material as hereinafter specified.

Sand and Water

2. All sand, or other materials used in lieu of sand, shall be hard, clean, free from vegetable and animal matter, salt or alkali

Sand shall be either pit or fresh-water sand ; sea sand shall not be used. If demanded by the architect in charge, the sand is to be washed before use. The sand shall be screened to remove gravel ; not more than 10 per cent. by weight shall pass a Standard I.M.M. No. 80 sieve, and the proportions of different particle-sizes shall be well distributed.

In lieu of sand, approved crushed hard rock, slag, clean hard burnt old brick, clean clinker or coal-cinder of approved origin, or hard-burnt clay may be used. Where a mortar-mill is in use, and approved materials in lieu of sand are crushed in the mill in the process of mixing the mortar, these shall be crushed so that the whole of the final product will pass through an I.M.M. Standard No. 5 sieve. Previously crushed materials other than sand shall conform to the fineness requirements for sand.

Clear, fresh water only shall be used in mixing the mortar, free from any unusual proportions of dissolved salts or gases. Sea water or brackish water shall not be used.

Alternative Types

3. Lime mortar used in structures where high stresses may occur shall conform to one of the two following categories :—

I. Naturally Hydraulic.
II. Pozzolanic.

In I, the hydrated or slaked lime shall be prepared in the manner hereinafter specified from a quicklime conforming to the Specification for Hydraulic Lime, Classes C 2 or C 3.

In II, the lime putty used may be prepared either from a Class A, Class B, or Class C 1 lime in accordance with the Specification for Lime Putty for Coarse Stuff, Plasterers' Work, Class C 1 being treated as Class B, or from a mechanically hydrated lime in accordance with the Specification for Lime Putty from Hydrated Lime, Finishing Coat. In all such cases, a sufficiency of finely-divided approved pozzolanic material must be added, such that Standard test briquettes made up with one part by weight of the lime in question, and three parts by weight of standard sand, together with substantially the same proportion of the pozzolanic material as it is proposed to use in the mortar, when tested in accordance with Clause 4, Tensile Strength (Hydraulic Lime and Sand) in the Specification for Hydraulic Lime, in all other particulars, will show a minimum tensile strength of 50 lb. per sq. in. ; provided that if it be proposed to use a large proportion of the pozzolanic material to replace part of the sand normally used in mortars, part of the standard sand specified in the above test may be replaced by an equal weight of the pozzolanic material ; in such case, if one-third of the standard sand be thus replaced by an equal weight of the pozzolanic material, the tensile strength exhibited shall not be less than 75 lb. per sq. in. ; alternatively, if two-thirds of the standard sand be thus replaced by the pozzolanic material, the tensile strength exhibited shall not be less than 100 lb. per sq. in. In no case shall more than two-thirds of the standard sand be thus replaced in this test.

Slaking the Lime

4. Slaked lime is to be prepared from hydraulic lime by sprinkling the requisite quantity of water on the lime, covering up lightly so as to retain the heat and moisture for 24 hours; and then thoroughly stirring and incorporating the sand, together with any more water required. Unless subsequently broken up in a mortar-mill, any lumps that will not pass through a I.M.M. Standard No. 5 sieve shall be removed by screening before use.

Gauging

5. All mortar shall be thoroughly mixed to a uniform consistence with only sufficient water to attain a plastic condition suitable for trowelling; and hydraulic mortar shall be used within 10 days of slaking the lime.

Proportions

6. In structures below the ground or water-level that are subject to load, not more than one part by volume of sand shall be used with one part by volume of hydraulic lime.

In structures above ground subject to load, the proportion of sand may be two parts by volume to one of hydraulic lime.

In structures above ground not subject to load, and in works of less importance, such as one-storey partition walls, not exceeding three parts by volume of sand may be used with one part of hydraulic lime.

Provided that where it is desired to substitute the hydraulic lime by a mixture of non-hydraulic lime putty or hydraulic lime with a pozzolana, the proportions shall be agreed upon previously by architect and builder, these depending on the efficiency of the pozzolana used.

Use of Pozzolanic Lime Mortars

7. Of pozzolanic mortars no more shall be mixed than can be used within 24 hours of mixing, and after hardening has once set in such mortars shall not be used.

Notes

Class A limes (*e.g.*, white chalk, mountain, " Buxton," etc.) should not be used in any exacting masonry work without pozzolanic additions.

Above ground level, Class C 1 (feebly hydraulic) limes may be used alone in positions where high stresses, particularly tensile or shearing stresses, are unlikely, However, Classes C 2 and C 3 limes, or pozzolanic mixtures, are preferable here.

For exacting positions and below ground level or water level, either a Class C 2 or C 3 hydraulic lime or a pozzolanic mixture should be specified, unless indeed a lime-cement mixture is adopted here. The pozzolanic mortar may be made with either a Class A or B, or Class C 1 (feebly hydraulic) lime, approved pozzolanic additions being made to these.

Slag, clinker, or coal-cinder, etc., used in lieu of sand, after crushing, should be viewed with suspicion if either a high sulphur content is shown on analysis, or if much readily combustible matter is present (high " loss on ignition ").

Pargetting or Parging of Chimney-Flues

Notes.—A particularly plastic and adhesive mortar is required for plastering or lining chimney-flues ; and one that resists effectively the large changes in temperature experienced. Tradition, which assigns to this process a term which is a debased usage of the name of the old artistic external rendering and decoration of timbered housewalls, " pargetry," also hands down to us the curious recipê containing cow-dung.

Doubtless there is some particular virtue in the organic constituents (largely colloidal) present, in furthering the requisite plastic qualities ; whilst vegetable fibrous matter also present will have a reinforcing effect. It is a matter of history that this ingredient has been used throughout the world as a cementitious and plastic finishing material from the earliest days ; the problem awaits scientific solution as to the particular physical or chemical actions implied, and as to a suitable synthetic substitute for the natural organic material.

Clause.—Pargetting or Parging of Chimney-Flues.

Mortar for pargetting or parging of chimney-flues shall be mixed with one part by volume of lime putty or hydrated lime, and three parts by volume of fresh cow-dung, free from litter, with sufficient water to produce a plastic mixture; and applied immediately. The lime putty or hydrated lime shall conform to the specifications for plastering limes.

References

—— *Amer. Soc. Testing Material Proc.*, 1922, **22**, and 1924, **24**. Tentative Specifications and Definitions.

—— *Brit. Eng. Standards Assoc.* Brit. Standard Specification for Portland Cement, 1925.

—— *Deutscher Verband fuer die Materialpruefung der Technik.* Bauing, 1925, **6**, 37, 66 (Provisional Specification for Trass, Clause 6, normal lime powder and normal lime putty).

DIBDIN, W. J., *Roy. Inst. Brit. Arch. Report*, 1911, loc. cit.

ECKEL, E. C., *Cements, Limes and Plasters*, loc. cit. (definition of hydraulic limes, p. 176 ; strength of hydraulic mortars, p. 187).

KIEPENHAUER, L., *Kalk und Moertel*, loc. cit.

KNIBBS, N. V. S., *Lime and Magnesia*, loc. cit. (Part I, Chapter IX ; Part III, Chapter III).

LONDON COUNTY COUNCIL. By-laws, 1891 (Oct. 13) ; Sect. 31, L.C.C. General Powers Act, 1890 ; Clause 1, Sect. 16. Metropolis Management and Buildings Act Amendment Act, 1878, Clause 3.

NAT. LIME ASSOC. (U.S.) *Pamphlets, Construction*, loc. cit., Chapters I, II, III, IVA, XVIIA, XVIII, Bull. 305A.

RIVINGTON, *Notes on Building Construction*, loc. cit. (7th edition, Part III, Chapter III).

SCHOCH, C., *Die Moderne Aufbereitung und Wertung der Moertel-Materialien.* Berlin, 1896 (p. 74).

—— *South African Eng. Standards Comm.* Lime, standardisation of methods of sampling and determination of available alkalinity, No. 3 (1913, Jan.) and suggested revision.

—— *Specification*, loc. cit.

—— *Tonind. Ztg.*, 1911, **35**, 1405 (Specification for lime. German Association of Lime Manufacturers).

—— *Tonind. Ztg.*, 1913, **36**, 417.

—— *Tonind. Ztg.*, 1925, **49**, 1427, 1439. (Russian standards).

U.S. Bureau of Standards, *Circular No.* 106, Lime, definitions and specifications. Tentative Specification for Masons' Hydrated Lime C9–19T (Nov., 1920), loc. cit. *Circular No.* 151, Wall Plaster (Jan., 1924), loc. cit. (Sand proportions, pp. 31–32.) *Circular No.* 201, Master Specification No. 250 for Quicklime for structural purposes (Feb., 1925). *Circular No.* 204, Master Specification No. 249, for Hydrated Lime for structural purposes (Mar., 1925). *Circular Nos.* 189, 203, 231, Lime Specifications for different purposes.

—— *U.S. Corps of Engineers, Professional Papers No.* 28 (1901). (Specification, Pozzolanic Cement.)

—— *Zement*, 1926, **15**, 71, *Technische Vorschriften fuer Bauleistungen* (Specification for Plaster-Work), loc. cit.

SECTION X.—BIBLIOGRAPHY

A LARGE number of literary references have been given at the ends of the respective sections, in connexion with the special topics dealt with therein. Amongst these it will be noticed that several works of reference frequently recur; these are certain textbooks and treatises which deal with the topics of lime and lime products in a wider manner, even if in some cases in conjunction with allied topics, such as cement, or general building construction.

For further information on the individual topics, the reader is referred to the specialised bibliographies mentioned; but for those who are interested in the subject in a general way, a brief résumé of the more accessible and instructive of the literary sources of information may be useful.

A general elementary treatment of the whole topic of lime mortars and plasters, their practical use in building, and some account of the classification of limes, from the builder's standpoint, will be found in the standard textbooks of building construction, and materials of construction, such as Mitchell, *Building Construction*, Advanced Course, 6th edition, London, 1909; and Rivington, *Notes on Building Construction*, Part III, 7th edition, London, 1919; several references to which have already been made. One or other of such textbooks should be available in any technical library.

A most comprehensive treatise on the chemistry, manufacture and uses of lime is that of Knibbs: *Lime and Magnesia*, London, 1924. This gives a brief history, together with an account of the classification and nomenclature of limestones and lime, and of the geological distribution of the former. A great number of data are given as to the chemical and physical properties of the oxide, hydrate, and carbonate; a description of the processes of quarrying and burning of lime; and some account of the varied uses of the product. A very large number of bibliographical references are also given, many to sources which would normally be somewhat inaccessible to the average student of the subject. The allied topic of magnesia is similarly treated in this volume. An excellent American work, covering a good deal of the same ground, and also giving a wealth of bibliographical references at the end of each chapter, is that of Eckel: *Cements, Limes and Plasters, their Materials, Manufacture, and Uses*, 2nd edition, London and New York, 1922. This treatise devotes a large proportion of space to the topics of Portland cement, gypsum plasters, magnesia cements, natural cements, slag cements, sand-lime bricks, &c., as well as a chapter on the manufacturing side of lime-burning, topics not dealt with in this Report; but the history, classification, occurrence, composition, and properties of lime and hydrated lime, pozzolana, and pozzolanic cements are dealt with in an exhaustive manner, European practice as well as American receiving due mention. Frequent references have been made to these two works.

The older works of Vitruvius, Vicat, Brunell, and others, to which reference has been made, are now more interesting from an historical standpoint than as a source of practical information; though of course Vicat's work forms the basis of the modern classification given in this Report. Those who are interested in the æsthetic side of plaster-work, and in the preparation of fine plaster surfaces for fresco and allied mural decorations, will find an immense fund of information and inspiration in Millar's large work on *Plastering*, 1897, and in the *Papers of the Society of Mural Decorators and Painters in Tempera*, 1925, **2** (loc. cit.).

For existing formal specifications of lime and lime products, both for use of the purchaser and for the architect in charge of works, reference should be made to the numerous tentative and final specifications in the *Proceedings of the American Society for Testing Materials*, particularly for 1924, and perhaps to *Specification*, Chatterton, London, 1926 (loc. cit.). Reference should also be made to the British Engineering Standards Association's Standard Specifications for comparable materials, such as Portland cement.

With regard to the strength of mortars, perhaps the most reliable and immediately accessible recent figures are those given in the *Report to the Royal Institute of British Architects*, by Dibdin, 1911, of which mention has already been made in its place.

A most excellent and comprehensive series of special Circulars and Technologic Papers has been issued in recent years by the American Bureau of Standards, extensive use of which has been made in the preparation of this Report. These are the Technologic Paper, No. 16, *The Manufacture of Lime*, 1913; Circular No. 30, *Lime, its Properties and Uses*, July, 1920; Circular No. 106, *Lime: Definitions and Specifications*, November, 1920; and Circular No. 151, *Wall Plaster*, January, 1924; the last being the *Report of the Bureau of Standards Plastering Conference* in 1923.

On the subject of the mechanical testing of mortars and cements, the effect of varying water-content, size of sand, &c., a vast literature exists in connexion with Portland cement, and almost any modern textbook on cement testing and allied topics will provide information applicable to lime mortars.

Most of the works mentioned here should be available in a good technical reference library of any pretensions; the American Bureau of Standards publications are available, *e.g.*, in the London Patent Office Library, or can be obtained from the Government Printing Office, Washington, U.S.A.

Introductory chapters on lime will be found in several textbooks which deal principally with the modern Portland cement, *e.g.*, Desch, *The Chemistry and Testing of Cement*, London, 1911 (loc. cit.); Doncaster, *Limes and Cements*, London, 1916; and Gillmore, *A Practical Treatise on Limes, Hydraulic Cements, and Mortars*, New York, 1896.

Of modern foreign textbooks, in addition to the classical treatise of Le Chatelier, *Recherches expérimentales sur la constitution des Mortiers Hydrauliques*, 2nd edit., Paris, 1904 (especially Part III, *Ciments et Chaux Hydrauliques*), there are comprehensive works such as Grange, *Chaux et Sels de Chaux*, Paris, 1894; Kiepenhauer, *Kalk und Mörtel*, Köln, 1907 (loc. cit.); and Schoch, *Die moderne Aufbereitung und Wertung der Mörtelmaterialien*, Berlin, 1896 (loc. cit.); and the two treatises of Burchartz, H., *Luftkalke und Luftkalkmörtel*, Berlin, 1908, and *Hydraulische Kalke*, Berlin, 1912 (loc. cit.).

Bibliographical lists of books and articles relating to lime are to be found in the *Journal of the Special Libraries Association* of New York, Vol. 12, No. 4 (April, 1921), with several hundred entries; on lime burning in Block, B, *Das Kalkbrennen*, Leipzig, 1924 (loc. cit.); a short list on lime and cements in general in Lesley, R. W., *History of Portland Cement Industry in U.S.A.*, Chicago, New York and London, 1924; a number of historical references in Quietmeyer, *Zur Geschichte der Erfindung des Portland-zementes*, Berlin, 1912; and a bibliography of Trass in *Zement*, 1923, **12**, 57, 63, Cantzler and Splittgerber. Specialized bibliographies of local mineral resources, such as the U.S.A. Geological Survey annual publications, are not of great interest to the general student of the subject.

Of *periodical* literature, as well as such British publications as the *Journal of the Royal Institute of British Architects*; the *Architects' Journal*; the *Builder*; *Building*; the *Illustrated Carpenter and Builder*; the *National Builder*; the *Industrial Chemist*; *Chemistry and Industry*; the *Abstracts* of the British and American *Chemical*, and *Ceramic Societies*, and *Science Abstracts*: some foreign technological papers frequently contain articles of interest to the student of lime, e.g. *Ciment* (Paris); *Revue des matériaux de construction et de travaux publics* (Paris); *Bauingenieur* (Berlin); *Tonindustrie Zeitung* (Berlin); *Zement* (Charlottenburg); *Chemical and Metallurgical Engineering* (New York); *Industrial and Engineering Chemistry* (Washington); and particularly *Rock Products* (Chicago). Considerable use has been made of the files of these journals in the preparation of this Report; some at least will be found in any well-equipped technical library, as in the Patent Office Library, London, and are invaluable as an indication of the most recent advances in the technology of lime and lime products.

SHORT GLOSSARY OF TERMS

N.B.—For a large number of technical terms, not defined here, references to the text will be found in the Index.

Air-Separated: Submitted to a procedure in which the fine and coarse particles are separated (and collected in distinct portions) by the regulated action of a current of air.

Air-Slaked Lime: Lime which has been hydrated and then partly carbonated by exposure to the moisture and carbon dioxide present in ordinary air. It consists therefore of slaked lime mixed with a varying proportion of inert calcium carbonate, *i.e.*, practically powdered limestone or chalk; and is useless for building purposes.

Argillaceous Matter: Clay substance, *i.e.*, essentially hydrated aluminium silicate, with varying amounts of lime, magnesia, iron oxide, titanium oxide, alkalies. &c.

Calcimine (*Kalsomine*)*:* Whitewash or coloured distemper.

Casein: An organic substance prepared from separated milk, which has the power of dissolving in and combining with solutions of alkalies to form glue-like substances that harden when exposed to air.

Clunch: (*a*) An impure chalk-marl occurring in the cretaceous deposits. (*b*) A generic term amongst builders for all chalk rocks that can be utilised directly as building stones.

Colloidal State: Matter, when in such a finely divided state that it will pass through a filter-paper but not through parchment, exhibits peculiar properties, due largely to the great surface area of each particle compared with its mass, and is said to be in the colloidal state (*e.g.*, glue, gelatine). Such particles are sub-microscopic.

Dead-burnt (of lime, cements, &c.): Materials burnt at so high a temperature that, after cooling, they react with water either extremely slowly or not at all. Applied especially to magnesia and plaster of paris.

Dolomite: A calcareous rock with a high content of magnesium carbonate, approximating to the composition $CaCO_3$, $MgCO_3$, *i.e.*, with 45·65 per cent. magnesium carbonate present.

Dolomitisation: Partial replacement of the calcium in calcium carbonate, in limestone, by magnesium, producing a mixture of magnesium and calcium carbonates, approaching ultimately to the composition of dolomite, $CaCO_3$, $MgCO_3$.

Doubled-up Work (Plastering): Plaster-work carried out in two or more coats, in which a second coat is applied over a former coat before this has had time to dry out.

Fibrous Plaster ("*Stick and Rag*"): Pre-cast sections, generally ornamental and of calcined gypsum plaster (plaster of paris) reinforced by a backing of canvas and laths, which can be affixed dry by screws or nails.

Flocculation: The clotting or clumping together of fine dispersed particles (*e.g.*, colloidal particles) to form large particles.

Fresco-painting (traditional): Artistic decorative painting on plaster surfaces whilst still fresh ("al fresco") and moist, giving soft effects and permanence.

Hawk (plasterer's tool): A board with a handle below used for holding the material preparatory to depositing it upon the surface to be plastered.

Hawkboy (traditional): Plasterer's assistant.

Kankar: Indian hydraulic mortar, made with lime from argillaceous lime stones.

Laitance : The smooth skin formed during the setting on the surface of Portland cement or cement concrete, produced by the water carrying up the fine material to the surface. This must be removed when plaster is subsequently applied, in order to provide for a bond between the plaster and surface.

Marl : A soft calcareous rock containing generally appreciable quantities of clay and sand, made up largely of shells; used (particularly in the U.S.A.) for Portland cement manufacture.

Salamander : A portable fire-pot or heating furnace used to accelerate the drying of plaster in cold or wet weather.

Santorin (earth) : A pozzolanic sand exported from the Greek island of Santorin.

Selenitic Lime : A lime, usually hydraulic, to which a small proportion of calcined gypsum had been added. This was considered to result in increased strength.

Shale : An argillaceous type of rock, characterised by cleavage-planes, intermediate between clay and slate.

Sinter, to : To heat to the point of initial fusion of the (mixed) material ; for instance, in the manufacture of Portland cement.

Slag : A by-product of the metallurgical industries. After the metal has been extracted from the ore, the molten residue, which constitutes the slag, is run out ; in chemical composition this consists mainly of calcium silicates together with variable impurities depending on the ore and flux used. Some slags when finely granulated by sudden cooling show marked pozzolanic properties.

Stucco : A plaster used for the exterior walls or other external surfaces of any building or structure. The word is used here without regard to the composition of the material, but defining only its use and location of use. Modern stucco often consists of Portland cement mortar.

Stucco Duro : Historic hard lime-marble stucco.

Surkhi : Burnt clay, used as a pozzolanic addition to mortar in India.

Trass : A siliceous rock, of volcanic origin, showing marked pozzolanic properties, found particularly in the neighbourhood of Andernach on the Rhine, and in common use in hydraulic mortars since 1682 in Germany and Holland.

INDEX

Printed under the authority of HIS MAJESTY'S STATIONERY OFFICE
By HARRISON AND SONS, LTD., 44–47, St. Martin's Lane, London, W.C.2,
Printers in Ordinary to His Majesty.

(B 34/3671)T 15657—1229/1055 1000 5/27 H & S Ltd. Gp. 34